Spacer Engineered FinFET Architectures

Spacer Engineered FinFET Architectures
High-Performance Digital Circuit Applications

Brajesh Kumar Kaushik
Sudeb Dasgupta
Pankaj Kumar Pal

CRC Press
Taylor & Francis Group
Boca Raton London New York

CRC Press is an imprint of the
Taylor & Francis Group, an **informa** business

CRC Press
Taylor & Francis Group
6000 Broken Sound Parkway NW, Suite 300
Boca Raton, FL 33487-2742

First issued in paperback 2020

© 2017 by Taylor & Francis Group, LLC
CRC Press is an imprint of Taylor & Francis Group, an Informa business

No claim to original U.S. Government works

ISBN 13: 978-0-367-57355-3 (pbk)
ISBN 13: 978-1-4987-8359-0 (hbk)

Library of Congress Cataloging-in-Publication Data

Names: Dasgupta, Sudeb, author. | Kaushik, Brajesh Kumar, author. | Pal, Pankaj Kumar, author.
Title: Spacer engineered FinFET architectures : high-performance digital circuit applications / Sudeb Dasgupta, Brajesh Kumar Kaushik, Pankaj Kumar Pal.
Description: Boca Raton : Taylor & Francis, CRC Press, 2017. | Includes bibliographical references.
Identifiers: LCCN 2016051118 | ISBN 9781498783590 (hardback) | ISBN 9781315191089 (electronic)
Subjects: LCSH: Metal oxide semiconductor field-effect transistors. | Silicon-on-insulator technology.
Classification: LCC TK7871.95 .D37 2017 | DDC 621.3815/284--dc23
LC record available at https://lccn.loc.gov/2016051118

Visit the Taylor & Francis Web site at
http://www.taylorandfrancis.com

and the CRC Press Web site at
http://www.crcpress.com

Contents

Preface

A FTER THE ANNOUNCEMENT MADE by Intel® to use tri-gate transistors commercially in the 22 nm technology node, FinFET technology has emerged as a major milestone in the field of nanoelectronics. In 2013, Taiwan Semiconductor Manufacturing Company (TSMC) unveiled the 16 nm FinFET process, which is one of the most advanced semiconductor technologies. It is the first integrated technology platform that demonstrated a less than 20 nm node. In order to keep pace with Intel and TSMC, the 14 and 10 nm FinFET process nodes will rapidly emerge as preferred choice among other semiconductor industries or foundries in the near future. Similar to the problems faced by any new technology, FinFETs with a sub-20 nm feature size face several design challenges. Most of these challenges arise due to technological restrictions such as the formation of ultrasharp source/drain junctions and the usage of precisely controlled dopants that can degrade the short-channel characteristics. This necessitates the use of undoped underlap regions that increase the channel resistance and reduce the drive current. Consequently, the gate-source/drain barrier restricts the carriers to flow from the source to drain, even at higher gate/drain bias.

In the past decade, high permittivity (k) spacer materials have emerged as the key enabler in enhancing device performance that provide a strong field coupling between the gate and the undoped underlap region; hence, reduces the increased source/drain resistance. However, it has limited applicability in high-performance circuits. The limitations are imposed due to an exorbitant increase in the fringe capacitance, that in turn, worsens the dynamic circuit performance. The other two inherent challenges associated with FinFETs that limit their applicability in high-performance circuit applications, are the higher magnitude of parasitic resistances/capacitances (due to its three-dimensional [3D] nature) and fin width quantization. Therefore, digital circuit designers need to modify their designs taking into account these critical issues so as to improve the overall performance in terms of device/circuit parameters such as I_{ON}, I_{OFF}, noise-immunity, and switching speed. At the device level, several researchers have focused on the integration of high-k materials as a gate dielectric and/or spacers. The fringing field phenomenon through high-k gate dielectric has been studied by few researchers from the circuit perspectives. However, a comprehensive study of the impact of 3D fringing field, due to high-k spacers, on the device and circuit performances is still required. To the best of our knowledge, none of the research work has ever explored the direct impact of a fringing field in enhancing the dynamic circuit performance of high-k spacer devices.

Process variability has emerged as one of the major concerns in a sub-20 nm gate length. The random variability in device/circuit increases sharply with a reduced feature size that can fail out any design. The major sources that degrade the tolerance of a device/circuit are random discrete dopant (RDD) induced fluctuations, oxide roughness variations (TOX), and metal grain-dependent work-function variations (WFV). Researchers have also explored various physical configurations such as symmetric and asymmetric device architectures to alleviate device-circuit codesign so as to improve the overall performance. However, contradictory observations have been made with respect to device/circuit immunity against the random variations, which result in an ambiguity about the true applicability of the devices. Therefore, it is necessary to investigate novel device architectures, with their circuit/static random access memory (SRAM) suitability and tolerance to random statistical variations.

This book primarily focuses on the novel device architecture that intelligently uses the high-permittivity spacer targeting high-performance device-circuit codesign (from the device level to the SRAM perspective) and its immunity to random statistical variations. The organization of the chapters is as follows. Chapter 1 provides an overview of micro- and nano-electronics, conventional Complementary Metal Oxide Semiconductor (CMOS) technology, the scaling limits, and beyond- conventional devices such as FinFET/Nanowires. Chapter 2 specifically deals with trigate FinFET technology and underlap/overlap architectures; and provides a brief review of the advancements of high-permittivity materials as dielectrics or spacers. Furthermore, the technological restrictions of high-permittivity spacers that limit their usage in a circuit are also illustrated. An optimized dual-permittivity (k) spacer concept is introduced in Chapter 3. It describes the proposed symmetric and asymmetric architectures, their fabrication methodology, and superior ON- and OFF-state electrostatics over the conventional (single/low-k) FinFET structure as well as the purely high-k spacer underlap FinFET structure. From this, an important and novel observation is made, i.e. the conduction band energy (CBE) barrier (in the ON-state) directly under the gate increases; which remains the same in a high-k device, even though the inner spacer permittivity is substantially increased. This chapter also physically interprets the OFF-state electrostatics associated with a dual-k structure with an increase in the inner spacer (k) value. A higher spacer permittivity can be beneficial in enhancing the stability, but at the cost of a delayed performance. Therefore, most researchers design logic circuits with low-permittivity spacers. Chapter 4 describes the role of fringe capacitances associated with the proposed architecture and demonstrates the suitability of high-k spacer materials for high-performance logic circuits, which simultaneously improves noise margins and delay performances. The circuit performances are evaluated based on the static and dynamic characteristics of a CMOS inverter and a three-stage ring oscillator. Furthermore, a detailed comparative analysis is carried out to thoroughly investigate the various competing effects in symmetric and asymmetric devices that affect the overall device and circuit performance. Chapter 5 explores the possibility of symmetric and asymmetric dual-k architectures for augmenting the SRAM design metrics. Cell performance is evaluated based on Static Noise Margins (SNMs) (hold, read, and write), read/write access times, and total leakage power. We have also demonstrated the effect of underlap length and power supply scalability on dual-k–based circuits/SRAMs. Finally, Chapter 6 explores the tolerance of symmetric and asymmetric dual-k spacer architectures and their SRAM performance using random statistical variations and sensitivity to key structural parameters.

About the Authors

Brajesh Kumar Kaushik received Ph.D. degree in 2007 from Indian Institute of Technology Roorkee, India. He served Vinytics Peripherals Pvt. Ltd., Delhi, as Research and Development Engineer in microprocessor, microcontroller, and DSP (Digital Signal Processing) processor-based system design. He joined Department of Electronics and Communication Engineering, G.B. Pant Engineering College, Pauri Garhwal, Uttarakhand, India, as Lecturer in July, 1998, where later he served as Assistant Professor from May, 2005 to May, 2006 and Associate Professor from May, 2006 to December, 2009. He joined Department of Electronics and Communication Engineering, Indian Institute of Technology Roorkee as Assistant Professor in December, 2009; where since April, 2014, he is working as Associate Professor. He has extensively published in several national and international journals and conferences of repute. He has also authored/co-authored several books and book chapters. He is reviewer of many international journals belonging to various publications such as IEEE (Institute of Electrical and Electronics Engineers), IET (Institution of Engineering and Technology), Elsevier, Springer, Taylor and Francis, Emerald, ETRI (Electronics and Telecommunications Research Institute), and PIER (Progress in Electromagnetics Research). He has served as General Chair, Technical Chair, and Keynote Speaker of many reputed international and national conferences. Dr. Kaushik is *Senior Member* of IEEE and member of many expert committees constituted by government and non-government organizations. He holds the position of Editor and Editor-in-Chief of various journals in the field of VLSI (Very Large Scale Integration) and microelectronics such as *International Journal of VLSI Design & Communication Systems* (*VLSICS*), AIRCC (Academy & Industry

Research Collaboration Center) Publishing Corporation. He also holds the position of Editor of *Microelectronics Journal (MEJ)*, Elsevier Inc.; *Journal of Engineering, Design and Technology (JEDT)*, Emerald Group Publishing Limited; *Journal of Electrical and Electronics Engineering Research (JEEER)*; and Academic Journals. He has received many awards and recognitions from the International Biographical Center (IBC), Cambridge. His name has been listed in Marquis Who's Who in Science and Engineering® and Marquis Who's Who in the World®. Dr. Kaushik has been conferred with Distinguished Lecturer award of IEEE Electron Devices Society (EDS) to offer EDS Chapters with a list of quality lectures in his research domain. His research interests are in the areas of high-speed interconnects, low-power VLSI design, memory design, carbon nanotube-based designs, organic electronics, FinFET device circuit co-design, electronic design automation (EDA), spintronics-based devices, circuits, and computing, image processing, and optics and photonics based devices.

 **Dr. S. Dasgupta** is an Associate Professor in the Department of Electronics and Communication Engineering at the Indian Institute of Technology (IIT) Roorkee, India. He received his Ph.D. degree in electronics engineering from IIT-Banaras Hindu University, Varanasi, India, in 2000. During his Ph.D. work, he carried out research in the area of effects of ionizing radiation on MOSFETs. Subsequently, he became a member of the faculty in the Department of Electronics Engineering at the Indian School of Mines, Dhanbad, India. In 2006, he joined the Department of Electronics and Communication Engineering, IIT Roorkee, as an Assistant Professor. He has authored/coauthored more than 200 research papers in peer-reviewed international journals and conferences. He is a member of Institute of Electrical and Electronics Engineers (IEEE), Electron Devices Society (EDS), and Indian Society for Technical Education (ISTE) and an associate member of Institute of Nanotechnology, the United Kingdom. He has been a technical committee member of the International Conference on Micro-to-Nano since 2006; he has also been nominated as Marquis's Who's Who in Science in Engineering® (the United States) awarded by Marquis, 2006–2008, and has been acting as an expert member of The

Global Open University, the Netherlands. He was awarded with the Erasmus Mundus Fellowship of the European Union in 2010 to work in the area of Resource Description Framework (RDF) at Politecnico Di Torino, Italy. He was a recipient of the prestigious Indo-U.S. Science and Technology Forum (IUSSTF) to work in the area of SRAM testing at the University of Wisconsin–Madison, Wisconsin, in 2011–2012. He was also awarded with the Deutscher Akademischer Austauschdienst (DAAD) Fellowship to work on analog design using reconfigurable logic at Technische Universität (TU), Dresden, Germany, in 2013. His areas of interest include nanoelectronics, nanoscale Metal Oxide Semiconductor Field-Effect Transistor (MOSFET) modeling and simulation, design and development of low-power novel devices, FinFET-based memory design, emerging devices in analog design, and development of reconfigurable logic. He has guided 10 Ph.D. scholars. Currently, he is supervising six Ph.D. candidates. He has been nominated for Indian National Academy of Engineering (INAE), Young Engineer Award. Dr. Dasgupta was a reviewer for *IEEE Transactions on Electron Devices, IEEE Electron Device Letters, IEEE Transactions on Nanotechnology, Superlattice and Microstructures, International Journal of Electronics, Semiconductor Science and Technology, Nanotechnology, IEEE Transactions on Very Large Scale Integration (VLSI) Systems, Microelectronic Engineering,* and *Microelectronic Reliability,* among others. He is also a member of Technical Program Committee (TPC) for VLSI Design Conference 2016 as well as for International Symposium on VLSI Design and Test (VDAT)-2016 (IIT, Guwahati, India). He has also been a member of the technical committees of various international conferences. He has presented a tutorial at VDAT-2014 and VLSI Design Conference, Bangalore, India, in 2015, among many others.

Pankaj Kumar Pal (S'12, M'16) received a B.Tech. degree in electronics and communication engineering from Gurukul Kangri University, Haridwar, India, in 2008, an M.Tech. degree in VLSI design from the National Institute of Technology (NIT), Hamirpur, India, in 2010, and a Ph.D. degree in microelectronics and VLSI from Indian Institute of Technology (IIT) Roorkee, India, in 2016. He is currently working as an Assistant Professor in the Department of Electronics Engineering, NIT, Uttarakhand, Srinagar (Garhwal), India. His current research interests include novel MOS-based devices, FinFET parasitic extraction, semiconductor device modeling, and low-power SRAM memory design. Dr. Pal received the Director's Medal at NIT-Hamirpur for being the branch topper in M.Tech. for 2010.

CHAPTER 1

Introduction to Nanoelectronics

1.1 INTRODUCTION

Ever since the first integrated circuit (IC) (Kilby 1964) and the complementary MOS technology (Wanlass 1967) were demonstrated in 1958 and 1963, respectively, the aggressive downscaling has continued that has actually fuelled the rapid growth of micro/nanoelectronics industries. In April 1965, Gordon Moore predicted that the number of transistors in an IC would double and the manufacturing cost would be reduced by half in every one and half years (Moore 1965). This trend had set a baseline for International Technology Roadmap for Semiconductors (ITRS) to discover or manufacture smaller and faster transistors that enabled the inventions of countless novel applications available to the community such as high-speed microprocessors and compact mobile phones.

After Moore's prediction, the electronic devices have evolved rapidly in terms of size, cost, and performance. The advances in the development of micro/nanoelectronic materials and device architectures in the past few years have ushered a new era of devices and circuits. While some of these new technologies are still in the developmental stages, many of them are on the way to become mainstream workhorses for the coming few years. The current physical gate length of transistors used in high-performance integrated circuits is around 22 nm (Shor and Luria 2012) and will possibly go further down to 7–10 nm by mid-2017, according to the projections made in recent ITRS trends (ITRS 2013). However, there are numerous challenges ahead for the semiconductor industries in their effort to track Moore's law beyond the sub-20 nm nodes. The main challenges in this regime are twofold: (a) reduction of leakage currents and (b) reduction in device variability to increase the yield (Xiong 2002).

1.2 SCALING AND LIMITATIONS OF A CLASSICAL CMOS DEVICE

Reducing the device dimensions not only results in a higher packing density but also leads to faster switching speed, lower power consumption, and lower manufacturing cost. Therefore in 1974, Dennard et al. (1974) reported the constant field scaling method to shrink the structural parameters of a device (horizontally as well as vertically) by a constant factor. However, as the gate length is shortened to enhance the operating speed and the chip density, the so-called short-channel effects (SCEs) arise. Therefore, the scaling is limited by various physical/electrical parameters such as V_{th} roll-off, drain-induced barrier lowering (DIBL), velocity saturation, and subthreshold swing (SS) degradation (Taur and Ning 2009). These effects have started plaguing the device characteristics mainly because of the reduced gate electrostatic control over the channel.

In general, an MOS transistor is said to be a short-channel device when the effective channel length is comparable to the sum of the source/drain junction depletion-layer widths (Tsuchiya et al. 1998). In short-channel MOSFETs, the electric-field lines that originate from the source/drain regions strongly influence the channel potential and govern the barrier (Nguyen and Plummer 1981). Ideally, the barrier was controlled by the applied gate field. To enhance the gate control over the channel, the gate dielectric layer must be made thinner and channel doping must be increased as suggested by the classical scaling rules (Dennard et al. 1974). Although this approach has been followed over the decades, in recent years this has given rise to a series of undesirable effects such as higher parasitic, increased gate tunneling current, mobility degradation, and random statistical variations such as random dopant fluctuation (RDF).

When the gate oxide thickness is reduced below 2–3 nm, the gate direct tunneling current increases exponentially, which in turn, increases the standby power dissipation. In sub-100 nm technology nodes, the gate oxide thickness has been reduced to the point where the power dissipated due to gate leakages is equivalent to the power for switching the circuit (Kim et al. 2003). Moreover, in highly doped MOSFETs, the existence of a large number of dopant ions obstructs the carrier motion because of Coulomb scattering; hence, the mobility degrades (Kittle 1976). Additionally, higher channel doping increases the surface electric field for a given inversion level, which results in reduced carrier mobility due to surface scattering (Takagi et al. 1994). The high surface electric field confines the carriers in a narrow potential well that results in quantum confinement effects (Stern and Howard 1967). Furthermore, a high gate oxide field depletes the poly-silicon gate with an appreciable amount of potential drop, thus reducing the effective gate bias (Lu et al. 1989). Quantum confinement (Vasileska et al. 1997; Taur and Ning 2009) and polysilicon depletion (Arora et al. 1995) lead to a threshold voltage shift and decrease the overall gate capacitance. On the other hand, RDF originates due to discrete dopant ions in the channel region. This effect is more prominent at smaller geometry because the total numbers of dopant ions are very small, which results in large statistical fluctuations. RDF also alters the transistor properties, especially, threshold voltage and drive current (Asenov et al. 1998). High body doping increases the electric field in the reverse biased source/drain-to-body junction, which significantly enhances the junction band-to-band tunneling (BTBT) current (Taur et al. 1997).

1.3 POTENTIAL TECHNOLOGIES BEYOND CONVENTIONAL CMOS

To improve the short-channel characteristics in classical devices, several methods such as super-steep retrograde profile (Jacobs and Antoniadis 1995), source/drain extension region, and halo implants (Tomimatsu et al. 2009) were suggested. In an aggressively scaled MOS architecture, the S/D-to-bulk capacitances are becoming a major issue that can be significantly minimized using SOI (silicon-on-insulator) technology. In extremely-thin SOI (ETSOI) platforms (also called fully-depleted SOI [FDSOI] or ultra-thin body [UTB] SOI), the gate control over the channel is significantly enhanced in comparison to the planar bulk technology. It is primarily due to thinner silicon film (T_{fin}) than the channel depletion depth. However, the use of SOI as an alternative to bulk technology remained confined only to the specific sectors and applications. The majority of the commercial market uses bulk process, championed by the foundries, including Intel®, TSMC, UMC, and GlobalFoundries. An alternative way to improve the gate electrostatic control, various multigate architectures can be used such as double-gate, FinFET, Si-nanowire, and gate-all-around. Such architectures can be implemented on SOI and bulk substrate as well.

Recently, multigate FETs have been seen as a better alternative for pushing the CMOS scaling under sub-20 nm gate lengths (Frank et al. 2001). The primary advantage of the multigate MOSFET is the excellent control of SCEs (Haensch et al. 2006; Colinge 2008) without relying on channel doping that makes it potentially scalable to the end of the ITRS roadmap. Having more than one gate around the channel improves the electrostatic integrity (*EI*) that is the measure of electric field lines originating from the source/drain and influencing the channel region. Various alternative device structures with multiple gates have been proposed to replace the classical planar MOSFET and extend the channel length scalability into the sub-22 nm regime.

1.4 EVOLUTION OF NOVEL DEVICE STRUCTURES

In a continuous effort to increase the drive current and also to control SCEs, SOI transistors have evolved from classical, planar, single-gate devices into three-dimensional devices with a multigate structure (i.e., double-, triple-, or quadruple-gate). Multigate FETs offer certain advantages over the conventional single-gate MOSFETs.

One of the most important advantages is the excellent gate control over the electrostatic charges. This increased charge control in the channel translates into improved SCEs (Colinge 2004a). Since, the channel is controlled electrostatically by the gate from multiple sides, the channel is better controlled by the gate than the conventional transistor structure. Unwanted leakage components are reduced and a small transistor can be used to continue the cost reduction through miniaturization. Improved gate control also provides the lower output conductance. This provides greater voltage gain, which is beneficial to RF/analog circuits as well as to the noise tolerance of digital circuits.

Another distinct characteristic of MugFETs is the increased ON-current, and therefore, faster circuit speed (Colinge 2003). One of the main advantages of using multiple-gate device is the highly improved electrical characteristic in the subthreshold regime (Colinge 2004b). The DIBL characteristic of a fully depleted multiple-gate transistor is much improved over a normal single-gate (SG) MOS transistor. The volume inversion is a phenomenon observed only in multiple-gate architectures. A device is said to be operating in volume inversion, if there is a strong coupling between two conducting channels (Balestra et al. 1987). In multiple-gate devices, the use of a very thin film allows us to downscale the devices without the need of using high channel doping densities and gradients (Choi et al. 2000). In fact, undoped films can be used; wherein, the fully depleted thin film prevents the punch-through mechanism. Besides this, the absence of dopant atoms in the channel increases the mobility by suppressing impurity scattering (Choi et al. 1995).

Multigate nanoscale devices have many advantages in circuit performance. A very high packaging density is possible because of the smaller size of these devices that have a short channel and thin film. Because of the higher mobility, transconductance can be higher; which gives more current gain and allows a higher operating frequency. Therefore, multiple-gate nanoscale devices are a potential candidate for RF and microwave applications (Mohankumar et al. 2010; Sohn et al. 2012). The analog performance of these devices is also superior. Moreover, the voltage gain is much higher than the gain of conventional bulk MOSFETs, especially in a moderate inversion region. In this section, we present a brief overview of planar DGFET and FinFET technology.

1.4.1 Planar Double-Gate (DG) MOSFETs

The first article on the DG MOS transistor was published by Sekigawa and Hayashi in 1984. The device was called XMOS because of its cross section that looks similar to the Greek letter Ξ (*Xi*). These double-gate transistors demonstrated a significant reduction of SCEs by the configuration; wherein, the drain-source channel was sandwiched between two independently fabricated gate/gate-oxide stacks as shown in Figure 1.1.

The double-gate FET can be thought of as an enhanced version of an FD SOI transistor with a very thin buried oxide (same thickness as the gate oxide). Only now, the back substrate is heavily doped and electrically connected to the top gate. As the top and bottom gates drive the substrate together, the gate-to-substrate coupling is perfect and the long channel SS is nearly 60 mV/decade. In addition, the control on SCEs is very good by virtue of a thin, fully depleted body and gate shielding of drain electric field lines from both sides. Because of the action of two

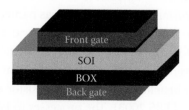

FIGURE 1.1   Planar double-gate MOSFET.

gates, the device can now be scaled to shorter gate lengths for the same body (and oxide) thickness.

Depending on the structure, materials, and applied gate voltages; DG MOSFETs may be categorized as symmetric or asymmetric. Symmetric DG MOSFETs are obtained when both the gates have the equal values of work function, oxide thickness, dielectric material, and input voltage applied to the gates. On the other hand, an asymmetric DG MOSFET is created by introducing asymmetry through input voltages, work functions, thicknesses, gate-dielectrics, materials, etc.

Planar DGFETs had been extensively researched during the initial phase of evolution of multigate transistors. Although the DG-MOS device offers significant advantages over SG devices, it has not played a significant role in the CMOS technology to date. The reason is that the planar DGFETs are difficult to fabricate (Nowak et al. 2004). There are problems in aligning the top and bottom gates as well as in building a low resistance contact at the bottom gate (Yin and Chan 2005). Manufacturing a self-aligned double-gate MOSFET has been the holy grail for device engineers and researchers ever since it was proposed. The mutual alignment of the top and bottom gates, and S/D diffusion is crucial; because any misalignment can result in parasitic capacitance. This problem is resolved in the FinFET, which has a self-aligned triple-gate structure.

1.4.2 FinFET Technology

The next major step forward in the electronics industry has been the introduction of FinFET technology. A FinFET is a new type of multigate 3D transistor that offers significant performance improvements and power reduction compared to existing planar CMOS devices. In a FinFET, the gate of the device wraps over the conducting drain-source channel as shown in Figure 1.2. This results in better electrical properties, providing lower threshold voltages and better performance as well as reductions in leakage and dynamic power.

In 1989, D. Hisamoto et al. fabricated the first self-aligned double-gate SOI structure. Initially, the transistor was named as DELTA, that is, fully **DE**pleted Lean-channel **TrA**nsistor. This was renamed as FinFET by the researchers of University of California, Berkeley, in 1999 (Huang et al. 1999). ITRS considers it as the potential candidate to replace classical MOSFETs due to the benefits of multigate transistor and relatively easier fabrication technique (ITRS 2003). The processes to fabricate DG-FinFET device options can be found in the literature (Mathew et al. 2004;

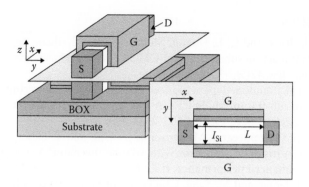

FIGURE 1.2   A cross-sectional view of the FinFET structure.

Masahara et al. 2006). Thereafter, extensive research has been carried out on FinFET/tri-gate technology from the device performance optimization and methodology to the circuit design.

In FinFET devices, the width of the silicon film (T_{fin}) is much smaller than its height (H_{fin}) that resembles the fin of a fish. The two side gates mostly control the device operation; consequently, an empirical scaling rule T_{fin} that defines the separation between the two side gates, must be less than one-third of channel length to suppress SCEs. As is evident, the electrical width of a triple-gate FinFET is $W = 2H_{fin} + T_{fin}$. In many cases, T_{fin} is small in order to have acceptably small SCEs. Moreover, in a DG-FinFET, the top gate is anyway ineffective. As a result, W is approximately $2H_{fin}$. As a result, the physics of a FinFET becomes largely similar to that of a DGFET. Thus, most of the literature that discuss compact model development for DGFETs, can be applied to FinFETs with a minor parameter (H_{fin}) adjustment.

A FinFET can be fabricated with its channel along different directions in a single die. Fabrication of planar MOSFET channels along any crystal plane other than ⟨100⟩ is difficult due to process variations and interface traps (Chang et al. 2004; Mishra and Jha 2010). However, the FinFETs can be fabricated along the ⟨110⟩ plane as well. This results in enhanced hole mobility. The ⟨110⟩ oriented FinFETs can be fabricated by simply rotating the transistor layout by 45° in the plane of a ⟨100⟩ wafer (Kang et al. 2010). Thus, the implementations of n-FinFETs along ⟨100⟩ and p-FinFETs along ⟨110⟩ lead to faster logic gates. These provide designers an opportunity to combat the inherent mobility difference between electrons and holes. However, this multiorientation scheme has an obvious drawback of an increased silicon area (Kang et al. 2010).

REFERENCES

Arora, N., E. Rios, and C. L. Huang, Modeling the polysilicon depletion effect and its impact on submicrometer CMOS circuit performance, *IEEE Trans. Electron Dev.*, 42(5), 935–943, 1995.

Asenov, A. et al., Random dopant induced threshold voltage lowering and fluctuations in sub-0.1 μm MOSFET's: A 3-D "atomistic" simulation study, *IEEE Trans. Electron Dev.*, 45(12), 2505–2513, 1998.

Balestra, F., S. Cristoloveanu, M. Benachir, J. Brini, and T. Elewa, Double-gate silicon-on-insulator transistor with volume inversion: A new device with greatly enhanced performance, *IEEE Electr. Device Lett.*, 8, 410–412, 1987.

Chang, L., M. Ieong, and M. Yang, CMOS circuit performance enhancement by surface orientation optimization, *IEEE Trans. Electron Dev.*, 51(10), 1621–1627, 2004.

Choi, J. H., Y. J. Park, and H. S. Min, Electron mobility behaviour in extremely thin SOI MOSFETs, *IEEE Electr. Device Lett.*, 16, 527–529, 1995.

Choi, Y. K., K. Asano, N. Lindert, V. Subramarian, T. J. King, J. Bokor, and C. Hu, Ultrathin-body SOI MOSFET for deep sub-tenth micron era, *IEEE Electr. Dev. Lett.*, 21, 254–255, 2000.

Colinge, J. P., Evolution of SOI MOSFETs: From single-gate to multiple-gates, *MRS Proceedings*, pp. 765, D1.6, 2003.

Colinge, J.-P., Multiple-gate SOI MOSFETs, *Solid-State Electron.*, 48(6), 897–905, 2004a.

Colinge, J.-P., *Silicon-on-Insulator Technology: Material to VLSI*, New York: Springer, 2004b.

Colinge, J.-P., *FinFETs and Other Multi-Gate Transistors*, New York: Springer-Verlag, 2008.

Dennard, R., F. Gaensslen, H. Yu, V. L. Rideout, E. Bassous, and A. LeBlanc, Design of ion-implanted MOSFETs with very small device dimensions, *IEEE J. Solidst Circ.*, 9(5), 256–68, 1974.

Frank, D., R. Dennard, E. Nowak, P. Solomon, Y. Taur, and H.-S. P. Wong, Device scaling limits of Si MOSFETs and their application dependencies, *Proc. IEEE*, 89(3), 259–288, 2001.

Haensch, W. et al., Silicon CMOS devices beyond scaling, *IBM J. Res. Dev.*, 50(4.5), 339–361, 2006.

Hisamoto, D., T. Kaga, Y. Kawamoto, and E. Takeda, A fully depleted lean-channel transistor (DELTA)—A novel vertical ultra thin SOI MOSFET, in *Electron Devices Meeting, 1989. IEDM'89. Technical Digest., International*, IEEE, Washington, DC, pp. 833–836, December 1989.

Huang, X., W. C. Lee, C. Kuo, D. Hisamoto, L. Chang, J. Kedzierski, E. Anderson, H. Takeuchi, Y. K. Choi, K. Asano, V. Subramanian, T. J. King, J. Bokor, and C. Hu, Sub-50 nm FinFET: PMOS, in *Electron Devices Meeting, 1999. IEDM'99. Technical Digest., International*, IEEE, San Francisco, USA, pp. 67–70, December 1999.

International Technology Roadmap for Semiconductor (ITRS). Available: http://public.itrs.net. SEMATECH, 2003.

International Technology Roadmap for Semiconductors. Available: http://public. itrs.net, 2013.

Jacobs, J. and D. Antoniadis, Channel profile engineering for MOSFET's with 100 nm channel lengths, *IEEE Trans. Electron Dev.*, 42(5), 870–875, 1995.

Kang, M., S. C. Song, S. H. Woo, H. K. Park, L. Ge, B. M. Han, J. Wang, G. Yeap, and S. O. Jung, FinFET SRAM optimization with fin thickness and surface orientation, *IEEE Trans. Electron Dev.*, 57(11), 2785–2793, 2010.

Kilby, J. S., Miniaturized electronic circuit, US patent 3138743, filed 1959, issued 1964.

Kim, N., T. Austin, D. Baauw, T. Mudge, K. Flautner, J. Hu, M. Irwin, M. Kandemir, and V. Narayanan, Leakage current: Moore's law meets static power, *Computer*, 36(12), 68–75, 2003.

Kittle, C., *Introduction to Solid State Physics*, New York: Wiley, 1976.

Lu, C. Y., J. Sung, H. Kirsch, S. Hillenius, T. Smith, and L. Manchanda, Anomalous C–V characteristics of implanted poly MOS structure in n+/p+ dual-gate CMOS technology, *IEEE Electr. Device Lett.*, 10(5), 192–194, 1989.

Masahara, M. et al., Demonstration of asymmetric gate oxide thickness 4-terminal FinFETs, in *Proceedings of the IEEE International SOI Conference*, Niagara Falls, New York, pp. 165–166, October 2006.

Mathew, L. et al., CMOS vertical multiple independent gate field effect transistor (MIGFET), in *Proceedings of the IEEE International SOI Conference*, Charleston, SC, pp. 187–189, October 2004.

Mishra, P. and N. K. Jha, Low-power FinFET circuit synthesis using surface orientation optimization, in *Proceedings of the Conference on Design, Automation and Test in Europe*, IEEE, Dresden, Germany, pp. 311–314, 2010.

Mohankumar, N., B. Syamal, and C. K. Sarkar, Influence of Channel and Gate Engineering on the Analog and RF Performance of DG MOSFETs, *IEEE Trans. Electron Devices*, 57(4), pp. 820–826, 2010.

Moore, G. E., Cramming more components onto integrated circuits, *Electronics Mag.*, 38, 1965.

Nguyen, T. and J. Plummer, Physical mechanisms responsible for short-channel effects in MOS devices, in *IEEE International Electron Devices Meeting Techical Digest*, Washington, DC, vol. 27, pp. 596–599, 1981.

Nowak, E. J., I. Aller, T. Ludwig, K. Kim, R. V. Joshi, C. T. Chuang, K. Bernstein, and R. Puri, Turning silicon on its edge [double gate CMOS/FinFET technology], *IEEE Circuits Device Mag.*, 20(1), 20–31, 2004.

Sekigawa, T. and Y. Hayashi, Calculated threshold-voltage characteristics of an XMOS transistor having an additional bottom gate, *Solid-State Electron.*, 27, 827, 1984.

Shor, J. and K. Luria, Evolution of thermal sensors in Intel processors from 90 nm to 22 nm, in *IEEE 27th Convention of Electrical and Electronics Engineers*, Israel, pp. 1–5, 2012.

Sohn, C., C. Kang, and R. Baek, Device design guidelines for nanoscale FinFETs in RF/analog applications, *IEEE Trans. Electron Dev.*, 33(9), 1234–1236, 2012.

Stern, F. and W. E. Howard, Properties of semiconductor surface inversion layers in the electric quantum limit, *Phys. Rev.*, 163, pp. 816–835, 1967.

Takagi, S., A. Toriumi, M. Iwase, and H. Tango, On the universality of inversion layer mobility in Si MOSFET's: Part I-effects of substrate impurity concentration, *IEEE Trans. Electron Dev.*, 41(12), 2357–2362, 1994.

Taur, Y. and T. H. Ning, *Fundamentals of Modern VLSI Devices*, 2nd ed. Cambridge, MA: University Press, 2009.

Taur, Y., D. A. Buchanan, W. Chen, D. J. Frank, K. E. Ismail, L. O. Shih-Hsien, G. A. Sai-Halasz, R. G. Viswanathan, H. J. C. Wann, S. J. Wind, and H. S. Wong, CMOS scaling into the nanometer regime, *Proceedings of the IEEE*, 85(4), 486–503, 1997.

Tomimatsu, T. et al., Cost-effective 28-nm LSTP CMOS using gate-first metal gate/high-k technology, in *Symposium on VLSI Technology Digest of Technical*, IEEE, Kyoto, Japan, pp. 36–37, 2009.

Tsuchiya, T., Y. Sato, and M. Tomizawa, Three mechanisms determining short-channel effects in fully-depleted SOI MOSFET's, *IEEE Trans. Electron Dev.*, 45(5), 1116–1121, 1998.

Vasileska, D., D. K. Schroder, and D. K. Ferry, Scaled silicon MOSFETs: Degradation of the total gate capacitance, *IEEE Trans. Electron Dev.*, 44(4), 584–587, 1997.

Wanlass, F. M., Low stand-by power complementary filed effect circuitry, US patent 3356858, filed 1963, issued 1967.

Xiong, W., K. Ramkumar, S. J. Jamg, J. T. Park, and J. P. Colinge, Self-aligned ground-plane FDSOI MOSFET, in *Proceedings of the IEEE International SOI Conference*, Williamsburg, VA, p. 23, 2002.

Yin, C. and P. C. H. Chan, Investigation of the source/drain asymmetric effects due to gate misalignment in planar double-gate MOSFETs, *IEEE Trans. Electron Dev.*, 52(1), 85–90, 2005.

CHAPTER 2

Tri-Gate FinFET Technology and Its Advancement

2.1 INTRODUCTION

In the past few decades, the semiconductor industry has grown consistently to meet the performance and computing requirements of various sectors ranging from medical applications to high-performance microprocessors. In recent years, the evolution of conventional silicon MOS-based integrated circuits (ICs) has made it possible to integrate greater functionality and complexity on a single chip. Today, ICs serve different needs ranging from low-power mobile applications, highly reliable military applications to high-performance computing applications. However, with the advent of new technology generations, there is a demand for increasing functionality within the same silicon area. This serves as a driving force towards the miniaturization of transistors.

Traditionally, bulk MOSFETs are used in digital ICs, and their scaling has been studied in detail. However, in the deep submicron regime, bulk MOSFETs are approaching the physical limits. With the reduction in geometric dimensions, devices are increasingly suffering from the short-channel effects (SCEs), high leakage power dissipation, increasing parasitic capacitances and resistances, large process variations, dominating interconnect delay, and so on. Most of these issues pertain to the inherent device structures. In conventional bulk MOSFETs, as channel length shrinks below 100 nm, channel doping engineering such as super halo (Taur and Nowak 1997) or asymmetric channel profile (Bansal and Roy 2005) is required to achieve the desired threshold voltage and to reduce SCEs. The high doping concentration increases the vertical electrical field resulting in mobility degradation and worsens subthreshold swing (SS). To improve the SS, the gate oxide thickness can also be reduced; which in turn, increases gate direct tunneling current. Also, as the dimensions reduce, dopant atoms are confined in a very small volume. Therefore, their placement in the channel region affects the device electrical characteristics. Because of process fluctuations, the placement of dopant atoms is random (Frank et al. 1999), resulting in a poor yield. Thus, bulk-MOSFETs are not suitable for extreme scaling. To overcome the bulk MOSFET scaling limitations, fully depleted thin-body silicon-on-insulator (FD-SOI) transistor structures have been proposed (ITRS 2013). Because of the thin body, a drain field is inhibited from penetrating deep into the channel. This gives more gate control over the channel resulting in improved SCEs and near-ideal subthreshold slope (Frank et al. 1992). Improved SCEs also relax the need of extreme scaling of gate dielectric thickness, thus reducing the gate direct tunneling leakage. To increase the drive current and to provide more gate control over the channel, several multigate structures such as planar double-gate (Wong et al. 1997), FinFET (Hisamoto et al. 2000), tri-gate (Chau et al. 2002), and omega-FET (Yang et al. 2002) have been proposed.

2.2 3D TRI-GATE/FinFET TECHNOLOGY

Among the multigate architectures, double-gate/tri-gate FinFETs have been proven to be a strong candidate for the future CMOS technology (Kuhn 2011). Therefore, Intel® already started its production at 22 and 14 nm technology nodes, as shown in Figure 2.1 (Auth et al. 2012).

The concept of the first planar double-gate transistor evolved in the late 1980s by Balestra et al. (1987) and Hisamoto et al. (1991). Besides the advantage of doubling the drive current by the presence of two channels, an additional interest existed in the possibility of volume inversion for thin-film devices. In volume inversion, the charge carriers are concentrated in the middle of the channel instead of SiO_2/Si interfaces, and this can be modulated by the front and back gate voltages (Colinge et al. 1990). Although the concept of the planar double-gate transistor appeared very promising, it was very challenging to fabricate such devices with perfectly aligned front and back gates. Early success was achieved by the

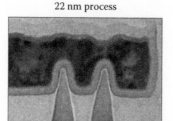

22 nm process   14 nm process

FIGURE 2.1   Microscopic images of Intel's® tri-gate FinFET design for 22 and 14 nm process nodes.

gate-all-around (GAA) transistors (Simoen et al. 1995). However, the main breakthrough of the double-gate transistors came in 1999, with the concept of the self-aligned FinFET. In this technology, the planar arrangement was abandoned for a vertical one and the top channel was replaced by two sidewall channels, wrapped around a silicon fin (Hisamoto et al. 2000; Kedzierski et al. 2001; Choi et al. 2002b).

FinFETs offer increased immunity to small-geometry effects, a near-ideal subthreshold slope, and certain other advantages such as increased mobility associated with lightly doped channel. Lower doping results in a weak electric field that further reduces the surface carrier scattering and gate tunneling. The use of an undoped or lightly doped body also provides immunity to threshold voltage and drive current variation due to statistical dopant fluctuations. For planar MOSFETs, the high substrate doping that used to control the SCEs also enabled threshold voltage adjustment. However, the freedom of threshold voltage adjustment was lost in the case of a FinFET due to the absence of channel doping. In FinFET, the required V_{th} was usually set by tunable work function metal gate (Liu et al. 2006). The undoped/lightly-doped channel increases the carrier mobility due to reduced Coulomb scattering. Thus, the FinFET architectures offer the potential for maintaining the scalability of the CMOS technology as it approaches the "end of the road-map" phase of its development (Kuhn 2011).

Depending on the type of substrate used, the FinFET can be broadly classified as a bulk or SOI type, as shown in Figure 2.2. Both types of FinFETs have merits associated with their structure. The main advantage

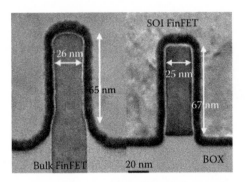

FIGURE 2.2   Microscopic images of fabricated bulk and SOI FinFETs.

of the bulk FinFET was its process compatibility with planar MOS technology and reduced self-heating. However, the SOI FinFETs are benefited from a lesser amount of junction capacitances. Apart from their merits/demerits, the choice between the bulk and SOI substrate is decided by the fabrication cost and ease of integration in the present technology setup. Typically, the body thickness is small compared to its height. Therefore, the two side gates have a prominent effect in controlling the channel inversion in comparison to the top gate. Also, the top gate influence on the channel reduces when top gate oxide is thicker than the side gate oxide. Since the FinFET is mainly controlled by two side gates, it is called double-gate (DG)-FinFET. Nevertheless, when the body thickness and top gate oxide are comparable to its height and side gate oxide, respectively, the presence of the top gate cannot be neglected. Such a device is called tri-gate FinFET (Kavalieros et al. 2006).

When the top gate is removed, the result is an independent-gate (IG) FinFET that can be controlled separately. IG FinFETs are primarily used to dynamically control the threshold voltage. DG-FinFETs can also be differentiated as symmetric or asymmetric based on dielectric material, thickness, and gate work function. Apart from the double-gate/tri-gate FinFETs, various other versions of the FinFET have been reported such as π-gate, Ω-gate, and quadruple-gate.

2.3 FinFET CLASSIFICATION

FinFETs have attracted increasing attention over the past one and a half decade because of the degrading short-channel characteristics of conventional MOSFETs (Reddy and Kumar 2005; Orouji and Kumar 2006; Kumar et al. 2014). It was the most researched device technology by the

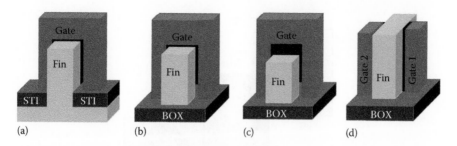

FIGURE 2.3 (a) Tri-gate bulk FinFET, (b) tri-gate SOI FinFET, (c) double-gate SOI FinFET, and (d) independent double-gate FinFET.

leading foundries/industries as well as academia (Sharma et al. 2009; Majumdar et al. 2010a,b). Depending on the usage at the device/circuit level, the FinFET structures can be classified as double-gate and tri-gate FinFETs, symmetric and asymmetric FinFETs, tied-gate and independent gate FinFETs, and bulk and SOI FinFETs (see Figure 2.3). This section will briefly discuss the various classifications and capabilities offered by the FinFET.

2.3.1 Bulk and SOI FinFETs

The FinFET can be made on bulk or SOI substrates known as the bulk FinFET (Figure 2.3a) or the SOI FinFET (Figure 2.3b–d), respectively (Kim and Lee 2005; Park et al. 2006; Kawasaki et al. 2006). However, the FinFETs implemented on SOI wafers are very popular and researched extensively. Unlike bulk FinFETs, where all fins share a common Si substrate (also known as the bulk), fins in SOI FinFETs are physically isolated. From the fabrication point of view, most of the foundries would prefer the bulk technology due to its easier migration from conventional bulk MOSFETs. However, FinFETs on both types of wafers are quite comparable in terms of cost, performance, and yield.

2.3.2 Double-Gate and Tri-Gate FinFETs

In general, FinFET can also be categorized as a tri-gate (Figure 2.3a,b) and double-gate FinFETs (Figure 2.3c,d). Both are the variant of a FinFET family. With an active third-gate on top of the Si-channel (as shown in Figure 2.3a,b), the FinFET architectures are abbreviated as tri-gate or triple-gate FinFETs. In triple-gate FinFETs, both the side surfaces and the top surface conduct current. The top gate also helps for self-alignment between the two side gates (Figure 2.4).

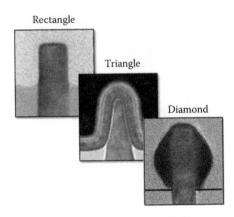

FIGURE 2.4   TEM crosssections of the fin shapes of different foundries.

In tri-gate FETs, the thickness of the dielectric on top of the fin is similar to that of the side-gate oxide thickness to activate the third gate. Due to the presence of third gate, the thickness of a fin also adds to channel width. Hence, tri-gate FinFETs enjoy slightly higher width advantage over the double-gate FinFETs. However, depending on the process flow of the respective foundries, the fin shape can be rectangular (Wu et al. 2010), triangular (Auth et al. 2012), or diamond.

Recently, Intel announced a big change in the electronic switches at the heart of its CPU. They introduced tri-gate FETs at the 22 nm node in the Ivy-Bridge processor in 2012 (Auth et al. 2012; Markoff 2011). Going forward, the firm announced commercial usage of three-dimensional transistors (triangular FinFET) instead of long-used planar MOS architecture because of their superior attributes at advanced technology nodes. In particular, FinFET demonstrates better performance in terms of leakage and dynamic power, intra-die variability, and retention voltage for static random-access memories (SRAMs).

For the double-gate FinFET, either the top-gate can be disabled by fabricating a thick hard-mask oxide layer (served as a tied double-gate FinFET) or etched out, and then, the two gates can be controlled independently. Therefore, the top surface of the fin does not conduct current in the double-gate FinFET. Yang et al. compared the double-gate FinFET with tri-gate FinFET and argued on its superiority over the tri-gate FETs in the long run (Yang and Fossum 2005). They showed that although the undoped tri-gate FETs enjoy more relaxed body thickness, they are not competitive with double-gate FinFETs in SCEs metrics. When trying to achieve comparable SCEs metrics, tri-gate FETs lose the scaling advantages and suffer from significant layout area disadvantages. However, with new triangular architecture, it is premature to declare a clear winner between double- and tri-gate FinFETs (Bhattacharya and Jha 2014).

2.3.3 Tied-Gate and Independent-Gate FinFETs

Based on the number of terminals, FinFETs can be categorized either as the tied-gate (TG) (Figure 2.3a–c) or IG (Figure 2.3d) configurations. TG FinFET is also known as shorted-gate (SG) or three-terminal (3T) FinFET, and the double-gate FinFET with four terminals (4T) is commonly known as IG FinFET. In TG configuration, the two side gates are connected through the top-gate to form a three-terminal device. Initially, the top gate was used for the perfect alignment of the two-side gates. Thus, in TG FinFETs, both gates are jointly used to control the electrostatics of the channel. Hence, TG FinFETs show higher ON-current (I_{ON}) compared to that of IG FinFETs. In IG configuration, the two side gates are separated and can be independently biased, which can be achieved by removing the top portion of the gate of a regular FinFET using chemical mechanical polishing (CMP).

With the four terminals, IG FinFET offers the flexibility of applying different voltages to its two gates; which enables the use of the back-gate bias to modulate V_{th}. This additional advantage of the IG FinFET proves it as a possible solution to the problem of width quantization. Over the past decade, IG-mode FinFETs have been extensively researched from circuit perspectives (Endo et al. 2008; Gupta et al. 2011). IG FinFETs suffer from high area penalty due to the need for placing two separate gate contacts and the comparatively higher parasitic (due to more interconnect lines). Several techniques to improve the circuit/SRAM performance metrics are discussed later in the section.

2.3.4 Symmetric and Asymmetric FinFETs

From device perspective, FinFETs can be further categorized as symmetric and asymmetric architectures. Symmetric FinFET architecture is perfectly symmetrical or similar with respect to the source/drain terminal as well as front and back gates. Most of the reported symmetric architectures that claiming better device/circuit performance, use some performance boosters such as high-k gate dielectric and spacer (Zhao et al. 2008). Rest of the symmetrical structure uses variations in the key process and structural parameters such as dopant segregation and S/D extension length.

Contrastingly, most of the asymmetric FinFET architectures targeted some specific device/circuit applications such as mitigation of the read/write conflict in the 6T SRAM cell.

Asymmetry in FinFET structures can be introduced in several ways. Initially in double-gate MOSFETs, the asymmetries were incorporated by applying different potential on both gates, using different work-function material gates, and by fabricating different front- and back-gate oxide material or thickness. However, over the past three to four years, several researchers have introduced asymmetries with respect to the source and drain terminal as well (Gupta et al. 2011; Moradi et al. 2011; Salahuddin et al. 2013). These types of asymmetries restrict the conventional inter-changeable source drain concept, but are useful for the applications having pass transistors. These asymmetric architectures are discussed in detail later in this book.

2.4 FinFET FABRICATION

To improve the short-channel characteristics of the FinFET, low fin thickness is required (Collaert et al. 2005). Typically, the fin thickness is equal to approximately half the gate length or lower that achieves good electrostatics in FinFETs (Tawfik and Kursun 2009). Since the minimum feature size associated with a technology is defined by lithography for the gate length, the fabrication of FinFETs requires sublithographic patterning technology for the formation of fins. Some examples of sublithographic patterning techniques include resist washing followed by oxide hard mask trimming (Asano et al. 2001) and spacer lithography (Choi et al. 2002b). There are several techniques for fabricating FinFETs, but only the spacer patterning technique has been discussed here.

In this subsection, we focus on spacer lithography and describe the features of this technique. Figure 2.5 shows the fabrication process for FinFETs using spacer patterning. A sacrificial layer of $Si_{0.4}Ge_{0.6}$ is deposited and patterned using lithography. Phosphosilicate glass (PSG) spacer is deposited around the SiGe sacrificial layer. The PSG layer is etched out from the top of SiGe so that the PSG layer remains only on the side of SiGe. Next, the SiGe sacrificial layer is etched out. The PSG layer that remains after the etching of the sacrificial layer serves as the mask for the formation of fins. Thus, the fin thickness is determined by the thickness of the PSG spacer layer and is not limited by lithography. In this manner, fins with low thickness can be formed by controlling the thickness of the PSG spacer layer deposited around the sacrificial layer.

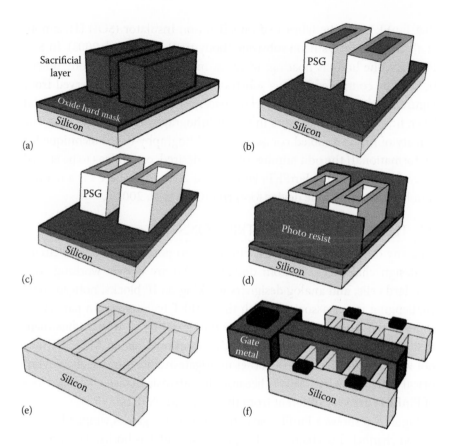

FIGURE 2.5 Fabrication of FinFETs using spacer patterning/lithography technique: (a) deposition and patterning of sacrificial $Si_{0.4}Ge_{0.6}$ layer, (b) deposition of phosphosilicate glass (PSG) spacer layer around $Si_{0.4}Ge_{0.6}$ and etching of PSG from the top of $Si_{0.4}Ge_{0.6}$, (c) etching of $Si_{0.4}Ge_{0.6}$, (d) formation of photo-resist mask for source/drain, (e) formation of fins and source/drain using PSG and photo-resist as masks, and (f) deposition of gate dielectric, gate metal and contacts and doping of source/drain to obtain the FinFET structure. (From Choi, Y. K. et al., *IEEE Trans. on Electron Dev.*, 49, 436–441, 2002a.)

After the etching of the sacrificial layer, a photo-resist is deposited and patterned for forming the raised source and drain (Choi et al. 2002a). Note that the raised source/drain (S/D) is often used for FinFET structures to reduce the S/D resistance (Vega and Liu 2009). With the photo-resist and the PSG spacers acting as the masks, fins with raised S/D are formed. The gate dielectric and gate metal/poly are deposited and patterned. This is followed by the formation of the gate spacers and doping of S/D. Finally, gate and S/D contacts are formed to obtain a FinFET. It may be mentioned that FinFETs can be fabricated on silicon-on-insulator (SOI) (Hisamoto et al. 2000) or on a silicon substrate (body-tied) (Park et al. 2003). In SOI FinFETs, the body is floating; whereas, in BT FinFETs, the body potential can be controlled using a substrate contact. It can be observed from Figure 2.5 that due to formation of spacer on both sides of the sacrificial layer, fins are formed in pairs in spacer lithography technology. Thus, the density of fins is doubled compared to a lithography-based technique for fin formation. If the odd number of fins is desired, one fin has to be etched away. In other words, FinFETs with $2i$ and $2i{-}1$ fins (where i is a natural number) have the same device footprint (Choi et al. 2002a).

2.5 TECHNOLOGICAL RESTRICTIONS

Like any other new technology, FinFETs also pose several device/circuit co-design challenges. Custom designers who are closely working with standard cells, and analog designers working on IP blocks, noticed some challenging issues associated with the FinFET technology. In particular, some of the design strategies such as the flexibility in width adjustment that have been used in the past for the conventional/planar MOSFET may not work for FinFETs and the other multigate architectures such as cylindrical gate/nanowires. This is because the intrinsic device characteristics of FinFETs are very different from the planar MOSFETs.

Figure 2.6 shows a FinFET structure in which a gate is wrapped on the silicon channel. The channel in the planar MOSFET is horizontal; whereas, the FinFET channel (also known as the fin) is vertical. Hence, the height

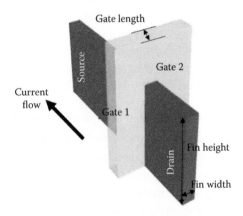

FIGURE 2.6 3D FinFET architecture showing the self-aligned gate wrapped on the Si-channel.

of the channel (H_{fin}) determines the width (W) of the FinFET. With planar transistors, standard cell designers can arbitrarily change transistor width to manage drive current. However, with FinFETs, designers cannot do the same. For this, designer can only add or subtract fins to change the drive current. In other words, the FinFET width must be a multiple of H_{fin}, that is, widths can be increased by using multiple fins. Also the fins come in discrete increments; thus, arbitrary FinFET widths are not possible, that is, we cannot add three-quarters of a fin. This issue is commonly known as "width quantization." Although smaller fin heights offer more flexibility, they lead to multiple fins; which in turn, lead to more silicon area. On the other hand, taller fins lead to less silicon footprint, but may also result in structural instability. Typically, the fin height is determined by the process engineers and is kept below four times the fin thickness (Collaert et al. 2005; Alioto et al. 2011).

There are several other issues also related to FinFETs that have not gained much attention as the width quantization has gained over the years. For example, body biasing will generally be impractical in FinFETs in comparison to the planar MOSFETs. Moreover, to modify the device threshold voltage by bulk biasing in 16/14 nm processes, one requires a very large voltage supply since the distance of the top of the fin (the active part) from the bulk is very large. This high supply voltage is neither feasible nor efficient enough in terms of power dissipation.

The long and narrow portion of the fin that is not under the gate is called the extension region. This region is technologically unavoidable, because it is not possible to have a steep lateral doping gradient, starting from a highly doped source/drain and ending at a lightly doped channel region. A lightly doped body is often preferred because it helps in reducing corner effects (Fossum et al. 2003b), random dopant fluctuations, and mobility degradation effects. As a result, FinFETs typically have a relatively large parasitic series resistance and capacitance. These parasitic resistance and capacitance represent another challenging area for the custom designer.

As the device shrinks further on the horizontal plane and the top gate dimension increases, a new coupling to neighboring elements appears and generates additional parasitic capacitances. Starting at 20 nm, C_{GS} (gate-to-source capacitance) and C_{GD} (gate-to-drain capacitance) effects become a larger concern and they contribute to the Miller effect that feeds the output of a circuit back into its input through the parasitic capacitances. Also, additional parasitic resistances in the source/drain area affect the device performance.

Clearly, it is observed that the introduction of FinFETs comes with several new design challenges. Most of these challenges are related to their effective usage while deriving maximum benefit of their characteristics. Nowadays, advanced TCAD/EDA tools can provide assistance in resolving these challenges and help in finding the best design using FinFETs.

2.6 REVIEW OF ADVANCEMENTS IN FinFETs

More recently, the FinFET concept has been translated to bulk Si substrates (Park et al. 2003), whereby the fins are now defined by shallow trench isolation (STI) regions. Bulk FinFETs allow fabrication on standard Si substrates without a major overhaul of fabrication line-up. Intel has adopted the 22 nm FinFET architecture (Jan et al. 2012) due to its combined advantages with scaling. Foundries such as TSMC followed suit for 16 nm technology node (www.eetimes.com/document.asp?doc_id=1319679), and other majors such as Samsung and GlobalFoundries collaborated to produce their 14 nm FinFETs (https://vrworld.com/2014/04/17/samsung-and-globalfoundries-collaborate-on-14nm-finfet/).

Over the past decade, the FinFET architecture has been the most researched and matured among all the multiple-gate technologies. Therefore, an extensive work has been carried out and reported on FinFET technology from the device performance optimization to the circuit/memory design perspectives. This section reviews the state-of-the-art FinFET device, circuit design, various optimization strategies, and performance metrics.

2.6.1 Device Structure and Performance Optimization

As the FinFET dimensions shrink, the SCEs and leakage currents continue to dominate, which degrades the device performance. Therefore, most of the recent research works have been focused on the device optimization considering various FinFET structures. Similar to the planar MOS devices, FinFETs also have overlap and underlap regions. The concept of "nonoverlapped" gate-source/drain (G-S/D) with low-doped channel was suggested (Trivedi et al. 2005) to facilitate the scaling of bulk-Si MOSFETs to 20 nm gate lengths. However, the difficulty in fabricating precisely controlled and well-defined doping profile necessitates the usage of undoped underlap regions. In the underlap structure, the effective channel length (L_{eff}) is significantly longer than the physical gate length (L_G) in weak inversion, while L_{eff} is comparable to L_G in strong inversion (Fossum et al. 2003a;

Yang et al. 2007). As a result, SCEs can be suppressed while maintaining the current driving capability. Previous studies have shown enhanced current drivability with an optimized G-S/D underlap structure (Shenoy and Saraswat 2003; Balasubramanian et al. 2003; Schulz et al. 2004). However, they have not considered the parasitic capacitance that strongly affects the switching speed in deeply scaled FinFETs.

Several researchers have focused on the integration of high-k materials to enhance the device performance (Agrawal and Fossum 2008, 2010; Shahrjerdi et al. 2009). The fringing-field phenomenon through these high-k gate dielectrics have been studied by few researchers from circuit perspectives (Mohapatra et al. 2003; Manoj and Rao 2007). A high-k gate dielectric could offer additional advantages such as significant enhancement in the performance and scalability (Vellianitis et al. 2007). The possible high-k advantages include thinner effective oxide thickness (EOT), which implies higher gate capacitance (C_{GG}) and ON-state current (I_{ON}). Also, the larger physical thickness (T_{hk}) of the high-k dielectric reduces the parasitic G-S/D outer-fringe capacitance (Kim et al. 2006; Manoj and Roa 2007). However, certain disadvantages come hand in hand with these merits. The larger T_{hk} results in the fringe-induced barrier lowering (FIBL) effect that significantly degrades the SCEs control (Mohapatra et al. 2003; Manoj and Roa 2007). Moreover, the channel mobility tends to degrade significantly due to poor quality of the Si/high-k dielectric interface (Zhu et al. 2004) and long-range scattering from optical phonons inherently present in high-k insulators (Fischetti et al. 2001). Furthermore, the integration of a high-k dielectric into the CMOS process presents alarmingly large technological challenges (Lee et al. 2005; Song et al. 2006). Manoj and Rao (2007) did a simulation-based study of the impact of high-k dielectrics on nanoscaled FinFET design and its performance. They compared $L_G = 32$ nm devices having different gate dielectric constants (k). They maintained the same OFF-state current (I_{OFF}) in each device by reducing T_{fin} with increasing k to suppress FIBL. They simulated CMOS ring oscillator, giving good insights, and reported a modest performance enhancement for an optimized permittivity of $k \approx 20$ (Agrawal and Fossum 2010).

The introduction of high-k spacers can provide a strong field coupling between the gate and undoped underlap region that reduces $R_{S/D}$ (Trivedi et al. 2005). This effect is commonly known as gate fringe-induced barrier lowering (GFIBL) that enhances the digital performance during a strong inversion regime (Sachid et al. 2008). Increasing S/D extension length (L_{EXT}) will increase the undoped/low-doped portion of L_{EXT} near the gate edge of the underlap FinFET. Therefore, restricting high-k dielectric to the gate sidewall can enhance the gate sidewall fringing fields; that in turn, can raise the barrier to conduction in the weak/moderate inversion regime. Ever since the devices have been scaled in the submicrometer regime, the parasitic capacitances have been a significant part of the gate capacitance that increases much faster as the scaling continues (Bansal et al. 2005). Trivedi et al. (2005) first reported the effect of the gate fringing field in the double-gate MOSFET on total gate capacitances using numerical simulation while discussing the effect of the abrupt and underlapped gate profile. Bansal et al. (2005) investigated the effect of the fringing-field component from the gate sidewall to the source through spacer in the double-gate MOSFET using conformal mapping. In 2010, Manoj et al. reported an enhanced fringe capacitance in FinFETs at 22 nm node compared to the equivalent planar MOSFETs. It is noted that high-k spacers increase the fringe capacitance (C_{fr}), which worsens the circuit delay in digital applications. To the best of our knowledge, none of the research works have ever explored the direct impact of the fringe field in enhancing the dynamic circuit performance in high-k spacer devices. Therefore, a comprehensive study of the impact of three-dimensional (3D) fringing field on the device and circuit level performances is still required.

2.6.2 Circuit Design Applications

This subsection focuses on the critical issues of FinFET circuit design. We review the merits and demerits of reported FinFET-based circuits in terms of the leakage power and functionality of analog/digital circuits and high-performance SRAM cells. Several authors investigated the potential of FinFET technology in digital circuits for high-performance digital applications (Bhoj and Jha 2013; Bhoj et al. 2013a,b) and SRAM memory design (Bansal et al. 2007). Regarding digital circuit design, most of the work has been based on independent gate configurations (Datta et al. 2007). It has been reported that an independent control of front and back gates (as dynamically adjustment of the V_{th}) can be exploited either to reduce standby power or to merge parallel transistors that reduces dynamic power through the reduction of parasitic capacitance. Cakici and Roy (2007) defined and presented different device as well as circuit design possibilities of DG-FinFETs. For example, Wu et al. (2006) and Nirmal et al. (2013) describe the suitability of double-gate MOSFETs in subthreshold circuits to achieve ultralow power consumption when speed is not of utmost importance. Chiang and Kim (2006) proposed a novel logic-circuit technique by employing an IG DGFET device. Using tied-gate configuration, Schmitt-Landsiedel and Werner (2009) showed the benefits of the multiple-gate inverter leading either to a lower leakage power at the same speed or to a higher speed at the same leakage as in the case of a conventional MOSFET. Independent-gate operation advantages in various circuits such as Schmitt triggers, dynamic logic circuits, sense amplifiers, and SRAM bit cells have been shown in Mahmoodi et al. (2004). Multithreshold-based FinFET sequential circuits with IG bias, work-function engineering, and gate-drain/source underlap engineering techniques are demonstrated in Tawfik and Kursun (2011). Lacord et al. (2012) compared planar and vertical FinFET structures based on propagation delays of inverter and NAND gate chains. Also, they investigated the impact of the width under several design rules for different FinFET configurations. Low-power multigate circuit design has been explored from the device/circuit point of view in Pacha et al. (2007). In Muttreja et al. (2007) and Rostami and Mohanram (2011), logic styles leveraging the tied and IG modes of FinFET operation have been investigated. FinFET latches and flip-flops have been studied in Tawfik and Kursun (2008). Due to small dimensions, FinFET is expected to suffer from the process and temperature variations. Metal-gate work-function variation is shown to be the most important contributor to the variation in V_{th} for FinFETs. FinFETs with asymmetric gate work functions in the form of n^+/p^+ polysilicon gates have been engineered and investigated in Kedzierski et al. (2001).

The FinFET offers lot of interesting device features that are potentially good for RF/analog applications. In the past half a decade, many research groups focused on FinFET applications in analog/RF circuit design. Wambacq and Verbruggen (2007) demonstrated the combination of a new gate stack in FinFET architecture that outperforms the comparable circuit realizations in planar bulk CMOS for low to moderate speed. In the microwave and millimeter-wave frequency region, planar bulk CMOS are still superior. The primary challenge for the FinFET structure in the coming years is the improvement of maximum cutoff frequency beyond 100 GHz. Fulde and Engelstädter (2007) demonstrated benefits of the FinFET in analog circuit applications and claimed that the introduction of novel gate stack materials (e.g., metal gate, high-k dielectric) and modified device architectures (e.g., fully depleted, undoped fins) can significantly affect the analog device properties. Also, the resulting benefits for speed, accuracy, and power trade-off in analog circuit design were presented. Thereafter, several device design engineering and optimization strategies were applied to FinFETs to enhance performance metrics in analog/RF domain. For example, Mohankumar et al. (2010) investigated the influence of channel and gate engineering on the analog/RF performance of DG FinFETs. Moreover, the analog and RF performance of a single halo double-gate MOSFET implemented with dual-material gate (DMG) technology has been investigated by Mohankumar et al. (2010). Recently, several publications have reported the analog/RF performance enhancement of underlap DGFETs with high-k spacers for low-power applications (Kranti and Armstrong 2007). Sohn et al. (2012) proposed some guidelines related to fin height and fin spacing for FinFET-based RF/analog applications.

2.6.3 Process Variations

Reduced feature size and limited photolithographic resolution cause statistical fluctuations in nanoscaled devices. These fluctuations cause variations in device as well as circuit performance parameters, such as V_{th}, I_{ON}, I_{OFF}, and Static Noise Margin (SNM). These process variations can be inter-die or intra-die, correlated or uncorrelated depending on the fabrication process. This leads to mismatched device strengths which degrades the yield of the die. That is why, continued scaling of planar MOSFETs has become so difficult. Therefore, Stojadinović and Ristic (1983), Stojadinović (1983), and Dimitrijev and Stojadinović (1987) reviewed the failure physics of ICs and their influence on device reliability in early days of scaling.

In planar MOSFETs, a sufficient number of dopants must be inserted in the channel to tackle the SCEs. However, this highly doped channel gives rise to random dopant fluctuations (RDFs) that further lead to significant variation in V_{th}. Since FinFETs enable better SCEs performance due to the presence of the second gate, they do not need a high channel doping to ensure a high V_{th}. Hence, designers have to keep the thin channel (fin) at nearly intrinsic levels (~10^{15} cm^{-3}). This reduces the statistical impact of RDFs on V_{th}. The desired V_{th} is obtained by engineering the work function of the gate material. The undoped/lightly doped channel also ensures better carrier mobility inside the channel. Thus, FinFET has emerged as superior to planar MOSFET by overcoming a major problem of process variations. However, due to its complicated structure and lithographic limitations, the FinFET does suffer from other process variations such as gate-edge roughness (GER), fin-edge roughness (FER), grain-dependent work-function variations (WFV), interface trap-charges fluctuations (ITC), gate oxide thickness, gate underlap, and positive/negative bias temperature instability (P/N-BTI) (Matsukawa et al. 2009; Mishra et al. 2010; Wang et al. 2011; Chaudhuri and Jha 2014). In sub-20 nm technology nodes, process variability has become one of the major concerns in the FinFET that can cause failure in any circuit/SRAM. Xiong and Bokor (2003) studied the sensitivity of performance parameters to various physical variations in devices designed with a nearly intrinsic channel. Choi et al. (2007) studied the temperature variations in FinFET-based logic circuits under key structural parametric variations. They showed that even under moderate process variations in gate length and body thickness (T_{fin}), more than 15% thermal runaway was possible in an IC, when primary input switching activity is 0.4. The effect of temperature variation is more severe in SOI FinFETs because the oxide layer under the fin has poor thermal conductivity. The heat generated within the fin cannot easily dissipate in SOI FinFETs. Bhoj et al. (2013a) evaluated the symmetric and asymmetric FinFETs under temperature variation and observed that the asymmetric tied gate FinFET remained the best and 100 times more advantageous than the symmetric one at higher temperature. They also plotted the distribution of I_{OFF} under process variations for the symmetric and asymmetric FinFETs. By optimization and/or modeling techniques, statistical variability in the SRAM cell has been carried out by many researchers (Kang et al. 2010; Ebrahimi and Rostami 2011; Hu et al. 2011).

2.7 FinFET DESIGN CHALLENGES AND ISSUES

Over the past few years, several active research areas have been explored such as device modeling (Mukhopadhyay et al. 2005; Prasad et al. 2015), and parasitic extraction (Bansal et al. 2004; Kumar et al. 2005, 2006; Bindu et al. 2007; Bhoj et al. 2013b). Most of the researchers have investigated the suitability of FinFETs in designing digital (Paul et al. 2006; Rasouli et al. 2009, 2010; Narang et al. 2012) and analog/RF circuits (Kolluri et al. 2007; Kundu et al. 2014; Ghosh et al. 2015; Koley et al. 2015). FinFET-based SRAMs have also been demonstrated by few authors (Ananthan and Roy 2006; Moradi et al. 2009; Bhoj and Jha 2014). Over the past three to four years, random/PVT variations in the FinFET have gained significant attention (Mahmoodi et al. 2005; Rao et al. 2010; Dadgour et al. 2010; Yang and Jha 2014). Despite the advantages of double/tri-gate structures, there are several challenges that need to be taken care of.

2.7.1 High-Permittivity Materials as Spacers

As the FinFET shrinks continuously, electrostatic control reduces giving rise to SCEs. Moreover, fabricating a precisely controlled and well-defined doping profile in sub-20 nm technology nodes is very difficult, which degrades the device performance. Therefore, the undoped underlap region is unavoidable in devices with gate length 20 nm or less (Yang et al. 2007). The underlap region helps in reducing SCEs but at the expense of drive current (I_{ON}). As the underlap length (L_{un}) further increases, the series resistance ($R_{S/D}$) starts dominating; and hence, the G-S/D barrier restricts the carriers to flow from source to drain, even at high bias. Introducing high-k spacers can provide a strong field coupling between the gate and the underlap region that reduces $R_{S/D}$ (Trivedi et al. 2005). However, it also increases the fringe capacitance (C_{fr}) that worsens the digital circuit performance in terms of delay and access time.

2.7.2 Parasitic Resistances and Capacitances

From a device point of view, one of the biggest challenges is the larger parasitic resistances and capacitances due to its own three-dimensional structure. For good short-channel control, a thin fin must be used. This results in a larger parasitic source/drain series resistance ($R_{S/D}$) due to the small cross-sectional area of the fin extension. To minimize $R_{S/D}$, a raised source/drain structure is often used. The raised source/drain is formed by a selective epitaxial growth process, which creates a nonrectangular raised source/drain cross section. Another important parasitic component is the outer fringe capacitances (C_{of}), which becomes significantly higher after raising the source and drain.

2.7.3 SRAM Design Challenges

A 6T SRAM cell has the most critical design consideration in terms of leakage power, delay, and noise margins. Over the past five to six years, the FinFET emerged as a very promising substitute to the conventional MOSFET for sub-22 nm technology nodes due to its excellent electrostatic control and reduced SCEs. Furthermore, it is possible to operate the FinFET at lower supply voltages because of lower (intrinsic) channel doping and larger effective channel width. Therefore, less dynamic power is achieved in SRAM circuits. Moreover, superior electrostatic control due to double/tri-gate also reduces the standby power consumption. Although the FinFET offers some advantages in SRAM design, it also poses several new challenges due to its novel 3D architecture. The most important and inherent design challenge associated with the FinFET is the width quantization effect that becomes more critical for circuits like SRAM; wherein, transistor sizing has a significant effect on its functionality. Moreover, the read/write conflict in the 6T SRAM cell aggravates the challenge posed by width quantization. Therefore, to design a power efficient, dense, and stable SRAM using FinFET architecture is a major concern nowadays.

2.8 SUMMARY

In this chapter, an existing literature review of double/tri-gate FinFET devices and their circuit/SRAM performances are discussed. This literature survey helped readers to identify various technical gaps in this area of research. Through the work presented in the subsequent chapters, an attempt has been made to bridge these technical gaps in order to have a better device in sub-22 nm technology nodes that fits well the requirements for high-performance circuit/memory applications.

REFERENCES

Agrawal, S. and J. G. Fossum, On the suitability of a high-*k* gate dielectric in nanoscale FinFET CMOS technology, *IEEE Trans. Electron Dev.*, 55(7), 1714–1719, 2008.

Agrawal, S. and J. G. Fossum, A physical model for fringe capacitance in double-gate MOSFETs with non-abrupt source/drain junctions and gate underlap, *IEEE Trans. Electron Dev.*, 57(5), 1069–1075, 2010.

Alioto, M. et al., Comparative evaluation of layout density in 3T, 4T, and MT FinFET standard cells, *IEEE Trans. VLSI Syst.*, 19(5), 751–762, 2011.

Ananthan, H. and K. Roy, Technology and circuit design considerations in quasi-planar double-gate SRAM, *IEEE Trans. Electron Dev.*, 53(2), 242–250, 2006.

Asano, K., Y. K. Choi, T. J. King, and C. Hu, Patterning sub-30-nm MOSFET gate with i-line lithography, *IEEE Trans. Electron Dev.*, 48(5), 1004–1006, 2001.

Auth, C. et al., A 22 nm high performance and low-power CMOS technology featuring fully-depleted tri-gate transistors, self-aligned contacts and high density MIM capacitors, in *Symposium on VLSI Technology Digest of Technical*, IEEE, Montgomery Village, pp. 131–132, 2012.

Balasubramanian, S., L. Chang, B. Nikolic, and T. J. King, Circuit-performance implications for double-gate MOSFET scaling below 25 nm, in *Proceedings of the Silicon Nanoelectronics Workshop*, Kyoto, Japan, pp. 16–17, June 2003.

Balestra, F., S. Cristoloveanu, M. Benachir, J. Brini, and T. Elewa, Double-gate silicon-on-insulator transistor with volume inversion: A new device with greatly enhanced performance, *IEEE Electron Device Lett.*, 8, 410–412, 1987.

Bansal, A., S. Mukhopadhyay, and K. Roy, Device-optimization technique for robust and low-power FinFET SRAM design in NanoScale era, *IEEE Trans. Electron Dev.*, 54(6), 1409–1419, 2007.

Bansal, A., B. Paul, and K. Roy, Impact of gate underlap on gate capacitance and gate tunneling current in 16 nm DGMOS devices, in *Proceedings of the IEEE SOI Conference*, South Corolina, pp. 94–95, October 2004.

Bansal, A., B. C. Paul, and K. Roy, Modeling and optimization of fringe capacitance of nanoscale DGMOS devices, *IEEE Trans. Electron Dev.*, 52(2), 256–262, 2005.

Bansal, A. and K. Roy, Asymmetric halo CMOSFET to reduce static power dissipation with improved performance, *IEEE Trans. Electron Dev.,* 52(3), 397–405, 2005.

Bhattacharya, D. and N. K. Jha, FinFETs: From devices to architectures, *Advances in Electro., Hindawi Publishing Corporation*, 2014, 1–21, 2014.

Bhoj, A. N. and N. K. Jha, Design of logic gates and flip-flops in high-performance FinFET technology, *IEEE Trans. VLSI. Syst.*, 21(11), 1975–1988, 2013.

Bhoj, A. N. and N. K. Jha, Parasitics-aware design of symmetric and asymmetric gate-workfunction finFET SRAMs, *IEEE Trans. VLSI. Syst.*, 22(3), 548–561, 2014.

Bhoj, A. N., R. V. Joshi, and N. K. Jha, Efficient methodologies for 3-D TCAD modeling of emerging devices and circuits, *IEEE Trans. Comput. Des. Integr. Circuits Syst.*, 32(1), 47–58, 2013a.

Bhoj, A. N., R. V. Joshi, and N. K. Jha, 3-D-TCAD-based parasitic capacitance extraction for emerging multigate devices and circuits, *IEEE Trans. VLSI. Syst.*, 21(11), 2094–2105, 2013b.

Bindu, B., N. Dasgupta, and A. Dasgupta, A unified model for gate capacitance—Voltage characteristics and extraction of parameters of Si/SiGe heterostructure pMOSFETs, *Energy*, 54(8), 1889–1896, 2007.

Cakici, R. T. and K. Roy, Analysis of options in double-gate MOS technology: A circuit perspective, *IEEE Trans. Electron Dev.*, 54(12), 3361–3368, 2007.

Chau, R. et al., Advanced depleted-substrate transistors: Single-gate, double-gate and tri-gate, in *Extended Abstracts of the International Conference on Solid-State Devices and Materials (SSDM)*, IEEE, Nagoya, Japan, pp. 68–69, 2002.

Chaudhuri, S. M. and N. K. Jha, 3D versus 2D device simulation of FinFET logic gates under PVT variations, *ACM J.E.T.C.*, 10(3), pp. 26:1–26.19, 2014.

Chiang, M. and K. Kim, High-density reduced-stack logic circuit techniques using independent-gate controlled double-gate devices, *IEEE Trans. Electron Dev.*, 53(9), 2370–2377, 2006.

Choi, J. H., J. Murthy, and K. Roy, The effect of process variation on device temperature in FinFET circuits, in *Proceedings of International Conference on Computer-Aided Design*, IEEE Press Piscataway, San Jose, CA, pp. 747–751, November 2007.

Choi, Y. K., T. J. King, and C. Hu, A spacer patterning technology for nanoscale CMOS, *IEEE Trans. on Electron Dev.*, 49(3), 436–441, 2002a.

Choi, Y. K., T. J. King, and C. Hu, Nanoscale CMOS spacer FinFET for the terabit era, *IEEE Electron Dev. Lett.*, 23(1), 25–27, 2002b.

Colinge, J. P., M. H. Gao, A. Rodriguez, H. Maes, and C. Claeys, Silicon-on-insulator gate-all-around device, in *IEDM Technical Digest*, New York: IEEE, p. 595, December 1990.

Collaert, N. et al., Tall triple-gate devices with TiN/HfO2 gate stack, in IEEE *Proceedings of Symposium on VLSI Technology*, Kyoto, Japan, pp. 108–109, June 2005.

Dadgour, H. F., K. Endo, V. K. De, and K. Banerjee, Grain-orientation induced work function variation in nanoscale metal-gate transistors-part I: Modeling, analysis, and experimental validation, *IEEE Trans. Electron Dev.*, 57(10), 2504–2514, 2010.

Datta, A., A. Goel, R. T. Cakici, H. Mahmoodi, D. Lekshmanan, and K. Roy, Modeling and circuit synthesis for independently controlled double gate FinFET devices, *IEEE Trans. Comput. Des. Integr. Circuits Syst.*, 26(11), 1957–1966, 2007.

Dimitrijev, S. and N. Stojadinović, Analysis of CMOS Transistor Instabilities, *Solid-State Electron.*, 30, 991–1003, 1987.

Ebrahimi, B. and M. Rostami, Statistical design optimization of finFET SRAM using back-gate voltage, *IEEE Trans. VLSI sys.*, 19(10), 1911–1916, 2011.

Endo, K. et al., Independent-gate four-terminal FinFET SRAM for drastic leakage current reduction, in *Integrated Circuit Design and Technology and Tutorial*, IEEE, Grenoble, France, pp. 63–66, 2008.

Fischetti, M. V., D. A. Neumayer, and E. A. Cartier, Effective electron mobility in Si inversion layers in metal-oxide-semiconductor systems with a high-k insulator: The role of remote phonon scattering, *J. Appl. Phys.*, 90(9), 4587–4608, 2001.

Fossum, J. G., M. M. Chowdhury, V. P. Trivedi, T.-J. King, Y. K. Choi, J. An, and B. Yu, Physical insights on design and modeling of nanoscale FinFETs, in *IEDM Technical Digest*, IEEE, Washington, DC, pp. 679–682, December 2003a.

Fossum, J. G., J. W. Yang, and V. P. Trivedi, Suppression of corner effects in triple gate MOSFETs, *IEEE Electron Device Lett.*, 24(12), 745–747, 2003b.

Frank, D. J., S. E. Laux, and M. V. Fischetti, Monte Carlo Simulation of 30 nm dual-gate MOSFET: How short can Si go?, in *IEDM Technical Digest*, International. IEEE, pp. 553–556, 1992.

Frank, D. J., Y. Taur, M. Ieong, and H. P. Wong, Monte Carlo modeling of threshold variation due to dopant fluctuations, in *Symposium on VLSI Technology*, Kyoto, Japan, pp. 169–170, June 1999.

Fulde, M. and J. Engelstädter, Analog circuits using FinFETs: Benefits in speed-accuracy-power trade-off and simulation of parasitic effects, *Adv. in Radio Sci.*, 5, 285–290, 2007.

Ghosh, S., K. Koley, and C. K. Sarkar, Impact of the lateral straggle on the analog and RF performance of TFET, *Microelectron. Reliab.*, 55(2), 326–331, 2015.

Gupta, S. K., S. P. Park, and K. Roy, Tri-mode independent-gate FinFETs for dynamic voltage/frequency scalable 6T SRAMs, *IEEE Trans. Electron Dev.*, 58(11), 3837–3846, 2011.

Hisamoto, D., T. Kaga, and E. Takeda, Impact of the vertical SOI 'DELTA' structure on planar device technology, *IEEE Trans. Electron Dev.*, 38(6), 1419–1424, 1991.

Hisamoto, D., W.-C. Lee, J. Kedzierski, H. Takeuchi, K. Asano, C. Kuo, E. Anderson, T.-J. King, J. Bokor, and C. Hu, FINFET-a self-aligned double-gate MOSFET scalable to 20 nm, *IEEE Trans. Electron Dev.*, 47(12), 2320–2325, 2000.

Hu, V. P., M. Fan, C. Hsieh, P. Su, and C. Chuang, FinFET SRAM cell optimization considering temporal variability due to NBTI/PBTI, surface orientation and various gate dielectrics, *IEEE Trans. Electron Dev.*, 58(3), 805–811, 2011.

International Technology Roadmap for Semiconductors (ITRS). Available: http://public.itrs.net, 2013.

Jan, C. H. et al., A 22 nm SoC platform technology featuring 3-D trigate and high-k/metal gate, optimized for ultra low power, high performance and high density SoC applications, in *IEDM Technical Digest*, IEEE, New York, p. 44, 2012.

Kang, M., S. C. Song, S. H. Woo, H. K. Park, L. Ge, B. M. Han, J. Wang, G. Yeap, and S. O. Jung, FinFET SRAM optimization with fin thickness and surface orientation, *IEEE Trans. Electron Dev.*, 57(11), 2785–2793, 2010.

Kavalieros, J. et al., Tri-gate transistor architecture with high-k gate dielectrics, metal gates and strain engineering, in *Symposium on VLSI Technology Digest of Technical*, Kyoto, Japan, pp. 50–51, 2006.

Kawasaki, H. et al., Embedded bulk FinFET SRAM cell technology with planar FET peripheral circuit for hp32 nm node and beyond, in *Proceedings of Symposium on VLSI Technology*, IEEE, Honolulu, pp. 70–71, June 2006.

Kedzierski, J. et al., High-performance symmetric-gate and CMOS-compatible Vt asymmetric-gate FinFET devices, in *IEDM Technical Digest,* IEEE, Washington, DC, pp. 492–495, December 2001.

Kim, S. H., J. G. Fossum, and J. Yang, Modeling and significance of fringe capacitance in nonclassical CMOS devices with gate-source/drain underlap, *IEEE Trans. Electron Dev.*, 53(9), 2143–2150, 2006.

Kim, S. Y. and J. H. Lee, Hot carrier-induced degradation in bulk FinFETs, *IEEE Electron Dev. Lett.*, 26(8), 566–568, 2005.

Koley, K., A. Dutta, S. K. Saha, and C. K. Sarkar, Analysis of high-κ spacer asymmetric underlap DG-MOSFET for SOC application, *IEEE Trans. Electron Dev.*, 62(3), 1–6, 2015.

Kolluri, S., K. Endo, E. Suzuki, and K. Banerjee, Modeling and analysis of self-heating in FinFET devices for improved circuit and EOS/ESD performance, in *IEDM Technical Digest*, IEEE, Japan, pp. 177–180, 2007.

Kranti, A. and G. A. Armstrong, Source/drain extension region engineering in nanoscale double gate MOSFETs for low-voltage analog applications, *Microelectron. Eng.*, 84(12), 2775–2784, 2007.

Kuhn, K., CMOS scaling for the 22 nm node and beyond: Device physics and technology, in *Symposium on VLSI Technology Digest of Technical*, IEEE, Kyoto, Japan, pp. 1–2, 2011.

Kumar, M. J., S. Gupta, and V. Venkataraman, Compact modeling of the effects of parasitic internal fringe capacitance on the threshold voltage of high-k gate-dielectric nanoscale SOI MOSFETs, *IEEE Trans. Electron Dev.*, 53(4), 706–711, 2006.

Kumar, M. J., V. Venkataraman, and S. K. Gupta, On the parasitic gate capacitance of small-geometry MOSFETs, *IEEE Trans. Electron Dev.*, 52(7), 1676–1677, 2005.

Kumar, S., V. Kumari, and M. Gupta, TCAD assessment of dual material gate nanoscale ringFET (DMG-RingFET) for analog and digital applications, in *Proceedings of International Conference on Devices, Circuits and Systems*, IEEE, Coimbatore, India, pp. 1–5, 2014.

Kundu, A., K. Koley, A. Dutta, and C. K. Sarkar, Impact of gate metal work-function engineering for enhancement of subthreshold analog/RF performance of underlap dual material gate DG-FET, *Microelectron. Reliab.*, 54(12), 2717–2722, 2014.

Lacord, J., J.-L. Huguenin, S. Monfray, R. Coquand, T. Skotnicki, G. Ghibaudo, and F. Boeuf, Comparative study of circuit perspectives for multi-gate structures at sub-10 nm node, *Solid-State Electron.*, 74, 25–31, 2012.

Lee, J. C. et al., Hafnium-based high-k dielectrics, in *Symposium on VLSI Technology Digest of Technical*, IEEE, Hisinchu, China, pp. 122–125, April 2005.

Liu, Y. et al., Investigation of the TiN gate electrode with tunable work function and its application for FinFET fabrication, *IEEE Trans. on Nanotechnology*, 5(6), 723–730, 2006.

Mahmoodi, H., S. Mukhopadhyay, and K. Roy, High performance and low power domino logic using independent gate control in double-gate SOI MOSFETs, in *Proceedings of the IEEE International SOI on Conference*, South Corolina, pp. 67–68, October 2004.

Mahmoodi, H., S. Mukhopadhyay, and K. Roy, Estimation of delay variations due to random-dopant fluctuations in nanoscale CMOS circuits, *IEEE J. Solid-St. Circ.*, 40(9), 1787–1795, 2005.

Majumdar, K., N. Bhat, P. Majhi, and R. Jammy, Effects of parasitics and interface traps on ballistic nanowire FET in the ultimate quantum capacitance limit, *IEEE Trans. Electron Dev.*, 57(9), 2264–2273, 2010a.

Majumdar, K., P. Majhi, N. Bhat, and R. Jammy, HFinFET: A scalable, high performance, low leakage hybrid n-channel FET, *IEEE Trans. Nanotechnol.*, 9(3), 342–344, 2010b.

Manoj, C. R. and V. R. Rao, Impact of high-k gate dielectrics on the device and circuit performance of nanoscale FinFETs, *IEEE Electron Dev. Lett.*, 28(4), 2007.

Manoj, C. R., A. B. Sachid, F. Yuan, C. Y. Chang, and V. R. Rao, Impact of fringe capacitance on the performance of nanoscale FinFETs, *IEEE Electron Dev. Lett.*, 31(1), 83–85, 2010.

Markoff, J., Intel increases transistor speed by building upward, May 2011 [Online]. Available http://www.nytimes.com/2011/05/05/science/05chip.html.

Matsukawa, T., S. O'uchi, and K. Endo, Comprehensive analysis of variability sources of FinFET characteristics, in *Proceedings of the Symposium on VLSI Technology*, Honolulu, HI: IEEE, pp. 118–119, June 2009.

Mishra, P., A. N. Bhoj, and N. K. Jha, Die-level leakage power analysis of FinFET circuits considering process variations, in *Proceedings of the 11th International Symposium on Quality Electronic Design*, California, pp. 347–355, 2010.

Mohapatra, N. R., M. P. Desai, and V. R. Rao, Detailed analysis of FIBL in MOS transistors with high-k gate dielectrics, in *Proceedings of the 16th Annual Conference VLSI Design*, New Delhi, India, pp. 99–104, January 2003.

Mohankumar, N., B. Syamal, and C. K. Sarkar, Investigation of novel attributes of single halo dual-material double gate MOSFETs for analog/RF applications, *Microelectron Reliab.*, 49(12), 1491–1497, 2009.

Mohankumar, N., B. Syamal, and C. K. Sarkar, Influence of channel and gate engineering on the analog and RF performance of DG MOSFETs, *IEEE Trans. Electron Dev.*, 57(4), 820–826, 2010.

Moradi, F., S. K. Gupta, G. Panagopoulos, D. T. Wisland, H. Mahmoodi, and K. Roy, Asymmetrically doped FinFETs for low-power robust SRAMs, *IEEE Trans. Electron Dev.*, 58(12), 4241–4249, 2011.

Moradi, F., D. T. Wisland, H. Mahmoodi, and T. V. Cao, Improved write margin 6T-SRAM for low supply voltage applications, in *Proceedings of the IEEE International SOC Conference*, Belfast, Northern Ireland, pp. 223–226, 2009.

Mukhopadhyay, S., K. Kim, C. Chuang, and K. Roy, Modeling and analysis of total leakage currents in nanoscale double gate devices and circuits, in *Proceedings of the International Symposium Low Power Electronics and Design*, IEEE, San Diego, pp. 8–13, August 2005.

Muttreja, A., N. Agarwal, and N. K. Jha, CMOS logic design with independent gate FinFETs, in *Proceedingsof the International Conference Computer Design*, Lake Tahoe, CA, pp. 560–567, October 2007.

Narang, R., M. Saxena, R. S. Gupta, and M. Gupta, Asymmetric gate oxide tunnel field effect transistor for improved circuit performance, in *International Conference On Devices, Circuits and System*, IEEE, Tamil Nadu, India, pp. 284–287, 2012.

Nirmal, D., P. Vijayakumar, D. M. Thomas, B. K. Jebalin, and N. Mohankumar, Subthreshold performance of gate engineered FinFET devices and circuit with high-k dielectrics, *Microelectron. Reliab.*, 53(3), 499–504, 2013.

Orouji, A. A. and M. J. Kumar, A new symmetrical double gate nanoscale MOSFET with asymmetrical side gates for electrically induced source/drain, *Microelectron. Eng.*, 83(3), 409–414, 2006.

Pacha, C. et al., Efficiency of low-power design techniques in multi-gate FET CMOS circuits, in *Proceedings of the European Solid-State Circuits Conference*, Munich, Germany, pp. 111–114, 2007.

Park, H. J. Cho, J. D. Choe, S. Y. Han, D. Park, K. Kim, E. Yoon, and J.-H. Lee Characteristics of the full CMOS SRAM cell using body-tied TG MOSFETs (Bulk FinFETS), *IEEE Trans. Electron Dev.*, 53(3), 481–487, 2006.

Park T., S. Choi, D. H. Lee, J. R. Yoo, B. C. Lee, J. Y. Kim, C. G. Lee, et al., Fabrication of body-tied FinFETs (Omega MOSFETs) using bulk Si wafers, in *IEEE Symposium on VLSI Technology*, IEEE, Kyoto, Japan, pp. 135–136, 2003.

Paul, B. C., A. Bansal, and K. Roy, Underlap DGMOS for digital-subthreshold operation, *IEEE Trans. Electron Dev.*, 53(4), 910–913, 2006.

Prasad, N., P. Sarangapani, K. Nadar, S. Nikhil, N. Dasgupta, and A. Dasgupta, An improved quasi-saturation and charge model for SOI-LDMOS transistors, *IEEE Trans. Electron Dev.*, 62(3), 919–926, 2015.

Rao, R., N. Dasgupta, and A. Dasgupta, Study of random dopant fluctuation effects in FD-SOI MOSFET using analytical threshold voltage model, *IEEE Trans. Device Mater. Reliab.*, 10(2), 247–253, 2010.

Rasouli, S. H., K. Endo, and K. Banerjee, Variability analysis of FinFET-based devices and circuits considering electrical confinement and width quantization, in *Proceedings of the 2009 International Conference on Computer-Aided Design*, IEEE, San Jose, CA, pp. 505–512, 2009.

Rasouli, S. H., H. F. Dadgour, K. Endo, H. Koike, and K. Banerjee, Design optimization of FinFET domino logic considering the width quantization property, *IEEE Trans. Electron Dev.*, 57(11), 2934–2943, 2010.

Reddy, G. V. and M. J. Kumar, A new dual-material double-gate (DMDG) nanoscale SOI MOSFET - Two-dimensional analytical modeling and simulation, *IEEE Trans. Nanotechnol.*, 4(2), 260–268, 2005.

Rostami, M. and K. Mohanram, Dual-Vth independent-gate FinFETs for low power logic circuits, *IEEE Trans. Comput. Aided Des.*, 30(3), 337–349, 2011.

Sachid, A. B., C. R. Manoj, D. K. Sharma, and V. R. Rao, Gate fringe-induced barrier lowering in underlap FinFET structures and its optimization, *IEEE Electron Device Lett.*, 29(1), 128–130, 2008.

Salahuddin, S. M., J. Hailong, and V. Kursun, Characterization of FinFET SRAM cells with asymmetrically gate underlapped bitline access transistors under process parameter fluctuations, in *IEEE International Conference on Electron Devices and Solid-State Circuits*, IEEE, Hong Kong, China, pp. 1–2, 2013.

Schmitt-Landsiedel, D. and C. Werner, Innovative devices for integrated circuits—A design perspective, *Solid-State Electron.*, 53(4), 411–417, 2009.

Schulz, T., C. Pacha, and L. Risch, Impact of technology parameters on device performance of UTB-SOI CMOS, *Solid State Electron.*, 48(4), 521–527, 2004.

Shahrjerdi, D., J. Nah, T. Akyol, M. Ramon, E. Tutuc, and S. K. Banerjee, Accurate inversion charge and mobility measurements in enhancement-mode GaAs field-effect transistors with high-k gate dielectrics, *Device Res. Conf.*, 29, 73–74, 2009.

Sharma, R. K., R. Gupta, M. Gupta, and R. S. Gupta, Dual-material double-gate SOI n-MOSFET: Gate misalignment analysis, *IEEE Trans. Electron Dev.*, 56(6), 1284–1291, 2009.

Shenoy, R. S. and K. C. Saraswat, Optimization of extrinsic source/drain resistance in ultrathin body double-gate FETs, *IEEE Trans. Nanotechnol.*, 2(4), 265–270, 2003.

Simoen, E., C. Claeys, S. Coenen, and M. Decreton, D.C. and low frequency noise characteristics of gamma-irradiated gate-all-around silicon-on-insulator MOS transistors, *Solid-State Electron.*, 38(1), 1–8, 1995.

Sohn, C., C. Kang, and R. Baek, Device design guidelines for nanoscale FinFETs in RF/analog applications, *IEEE Trans. Electron Dev.*, 33(9), 1234–1236, 2012.

Song, S. C. et al., Highly manufacturable advanced gate-stack technology for sub-45-nm self-aligned gate-first CMOSFETs, *IEEE Trans. Electron Dev.*, 53(5), 979–989, 2006.

Stojadinović, N., Failure physics of integrated circuits—A review, *Microelectron. Reliab.*, 23, 609–707, 1983.

Stojadinović, N. and S. Ristić, Failure physics of integrated circuits and relationship to reliability, *Physica Status Solidi (a)*, 75, 11–48, 1983.

Taur, Y. and E. J. Nowak, CMOS devices below 0.1 μm: How high will performance go?, *IEDM Tech. Dig.*, IEEE, USA, pp. 215–218, 1997.

Tawfik, S. A. and V. Kursun, Low-power and compact sequential circuits with independent-gate FinFETs, *IEEE Trans. Electron Dev.*, 55(1), 60–70, 2008.

Tawfik, S. A. and V. Kursun, Parameter space exploration for robust and high performance n-channel and p-channel symmetric double-gate FinFETs, in *1st Asia Symposium on Quality Electronic Design*, IEEE, Kuala Lumpur, Malaysia, pp. 246–251, July 2009.

Tawfik, S. and V. Kursun, Multi-threshold voltage FinFET sequential circuits, *IEEE Trans. VLSI,* 19(1), 138–143, 2011. Available online at: www.technavio.com/blog/samsung-globalfoundries-team-up-to-makefinfet-chips-challenge-tsmc.

Trivedi, V., J. G. Fossum, and M. M. Chowdhury, Nanoscale FinFETs with gate-source/drain underlap, *IEEE Trans. Electron Dev.,* 52(1), 56–62, 2005. Available online at: www.eetimes.com/document.asp?doc_id=1319679.

Vega, R. A. and T. J. K. Liu, Three-dimensional FinFET source/drain and contact design optimization study, *IEEE Trans. Electron Dev.,* 56(7), 1483–1492, 2009.

Vellianitis, G. et al., Gate stacks for scalable high-performance FinFETs, in *IEDM Technical Digest,* IEEE, Washington, DC, pp. 681–684, 2007.

Wambacq, P. and B. Verbruggen, The potential of FinFETs for analog and RF circuit applications, *IEEE Trans. Circuits Syst.,* 54(11), 2541–2551, 2007.

Wang, X., A. R. Brown, B. Cheng, and A. Asenov, Statistical variability and reliability in nanoscale FinFETs, in *IEDM Technical Digest,* Washington, DC, pp. 541–544, 2011.

Wong, H. S. P., K. K. Chan, and Y. Taur, Self-aligned (top and bottom) gate double-gate MOSFET with 25 nm thick silicon channel, in *IEDM Technical Digest,* USA, pp. 427–430, 1997.

Wu, C. C. et al., High performance 22/20 nm FinFET CMOS devices with advanced high-K/metal gate scheme, in *IEDM Technical Digest,* IEEE, California, pp. 27.1.1–27.1.4, 2010.

Wu, X., F. Wang, and Y. Xie, Analysis of subthreshold FinFET circuits for ultra-low power design, in *IEEE International SOC Conference,* Austin, TX, pp. 91–92, 2006.

Xiong, S. and J. Bokor, Sensitivity of double-gate and FinFET devices to process variations, *IEEE Trans. Electron Dev.,* 50(11), 2255–2261, 2003.

Yang, F. L., H.-Y. Chen, F.-C. Chen, C.-C. Huang, C.-Y. Chang, H.-K. Chiu, C.-C. Lee, et al., 25 nm CMOS Omega FETs, in *IEDM Technical Digest,* IEEE, California, 255–258, 2002.

Yang, J. W. and J. G. Fossum, On the feasibility of nanoscale triple-gate CMOS transistors, *IEEE Trans. Electron Dev.,* 52(6), 1159–1164, 2005.

Yang, J. W., P. M. Zeitzoff, and H. H. Tseng, Highly manufacturable double-gate finFET with gate-source/drain underlap, *IEEE Trans. Electron Dev.,* 54(6), 1464–1470, 2007.

Yang, Y. and N. K. Jha, FinPrin: FinFET logic circuit analysis and optimization under PVT variations, *IEEE Trans. Very Large Scale Integr. Syst.,* 22(12), 2462–2475, 2014.

Zhao, H., Y. C. Yeo, S. C. Rustagi, and G. S. Samudra, Analysis of the effects of fringing electric field on FinFET device performance and structural optimization using 3-D simulation, *IEEE Trans. Electron Dev.,* 55(5), 1177–1184, 2008.

Zhu, W., J. P. Han, and T. P. Ma, Mobility measurement and degradation mechanisms of MOSFETs made with ultrathin high-k dielectrics, *IEEE Trans. Electron Dev.,* 51(7), 98–105, 2004.

CHAPTER 3

Dual-*k* Spacer Device Architecture and Its Electrostatics

3.1 INTRODUCTION

As FinFET devices are drastically scaled, SCEs and leakage currents continue to dominate and affect device performance. Therefore, the large drive current projected by the ITRS for sub-20 nm FinFET devices has not been achieved yet. In the sub-20 nm regime, suppression of SCEs can be achieved by incorporating gate-source/drain (G-S/D) underlap regions. In an underlap structure, the dynamic effective channel length (L_{eff}) is significantly longer than the physical gate length (L_G) in weak inversion, while L_{eff} is comparable to L_G in strong inversion (Fossum et al. 2003). The underlap region helps in reducing SCEs, but at the expense of drive current (I_{ON}) (Yang et al. 2007). This is primarily because of the increased underlap resistance that starts dominating the total source/drain series resistance ($R_{S/D}$) with an increase in the underlap length (L_{un}). Consequently, the G-S/D barrier restricts the flow of carriers from source to drain, even at higher gate/drain bias.

Spacer engineering plays a significant role in describing the electrostatics and charge dynamics of the underlap devices. The undoped underlap region increases the total source/drain series resistance that in other words

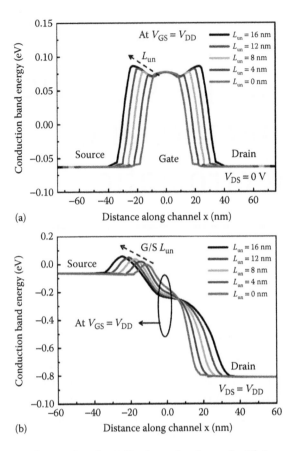

(a)

(b)

FIGURE 3.1 Conduction band profile along the channel with increasing underlap lengths (a) at $V_{DS} = 0$, and (b) $V_{DS} = V_{DD}$, when $V_{GS} = V_{DD}$.

creates a barrier (near the gate edges) in the OFF-state. This barrier is more prominent with an increase in underlap length, as shown in Figure 3.1a. With an increase in the drain potential, the underlap barrier on the drain side reduces without affecting much the source-side (G/S) underlap barrier, as shown in Figure 3.1b. In a large underlap device, this source-side underlap barrier is prominently high that actually restricts the carriers to flow from source to drain in the ON-state.

In the past decade, high-permittivity (k) spacer materials acted as a key enabler in enhancing the device performance. It provides a strong field coupling, that is, gate fringe-induced barrier lowering (GFIBL), between the gate and the undoped underlap region; hence, reduces the raised source/drain series resistance (Sachid et al. 2008). This results in a better gate control and improved device performance. However, replacing low-*k* spacer with a high-*k* spacer leads to problems related to trap charges and parasitic capacitance that significantly affect the device performance. In general, the high-permittivity spacer materials have limited applicability in high-performance digital circuits (Zhao 2008). This limitation is imposed due to an excessive increase in fringe capacitance (C_{fr}); that in turn, worsens the circuit delay/access time. The other problem is associated with interfacing of high-*k* material to silicon body which induces trapped charges that severely degrade the carrier mobility due to increased Coulomb scattering at the Si/spacer interface (Colinge 2008). As the *k* value increases, the mobility degradation will be much higher due to enhanced trap charges. Therefore, it is necessary to diligently use high-permittivity spacers so that the electric fringing field only converges toward the high energy barrier, which can help in reducing the extra $R_{S/D}$. This also helps in improving the device and circuit performance in the digital domain.

At the device level, several researchers have focused on the integration of high-permittivity materials as gate-oxide and/or spacers (Chen et al. 2005; Agrawal and Fossum 2008; Jan et al. 2012; Koley et al. 2015). The fringing field phenomenon through the high-*k* gate dielectric generating, FIBL, has been studied by few researchers from device and circuit perspectives (Mohapatra et al. 2002; Rao and Mohapatra 2004; Manoj and Rao 2007; Nirmal et al. 2013). There is a continuous reduction in dimensions of the device due to scaling that motivates a comprehensive study of a three-dimensional (3D) fringing field due to high-*k* spacers on both device and circuit performance. To the best of our knowledge, none of the research work has ever explored the direct impact of the 3D fringing field in enhancing the dynamic circuit performance using high-*k* spacers. Moreover, there are no studies that trace the electric flux across spacers of various dielectric constants.

This chapter presents a comprehensive analysis of the effect of high-permittivity spacers on underlap tri-gate FinFETs. This study helps the designer to make the effective use of such spacer-engineered devices for enhanced performance. Section 3.2 introduces the dual-permittivity (*k*) spacer concept within the device. Section 3.3 presents the optimization strategy for proposed symmetric (SymD-*k*) and asymmetric dual-*k* spacer (AsymD-k) tri-gate architectures and the simulation methodology adopted to explore the benefits of employing high-k spacers in symmetric and asymmetric devices. Section 3.4 provides the proposed methodology of fabricating symmetric and asymmetric dual-k spacer FinFETs. The superior ON-/OFF-state electrostatics and merits of dual-k FinFETs over the conventional (single/low-k spacer) as well as the purely high-k spacer underlap FinFET structure are discussed in Section 3.5. An extensive three-dimensional (3D) TCAD device simulation examines both ON- and OFF-state electrostatics with high-permittivity sidewall spacers to reveal the effects of the 3D fringing field phenomenon; that in turn, modulates charge dynamics inside the channel. Furthermore, this section also physically interprets the ON/OFF state electrostatics associated with the dual-k structure with an increase in the inner spacer k value. Moreover, Section 3.6 distinguishes the competing effects of symmetric and asymmetric dual-k spacer structures. This study helps in understanding the favorable field dynamics of their respective electrostatics and its influence on high-performance circuit applications. Finally, Section 3.7 briefly summarizes this chapter.

3.2 DUAL-k SPACER DEVICE ARCHITECTURE

The optimized dual-k spacers can effectively reduce the G-S/D underlap barriers by the concentrated fringing field. Apparently, the drain-side barrier in any underlap device inherently reduces by drain bias in the ON-state. However, the source-side underlap barrier is almost unaffected by drain bias; that in turn, restricts the carrier flow from source to drain (as shown in Figure 3.1b). Therefore, the optimum high-k spacer can be effectively used, either on both sides (abbreviated as SymD-k) to maintain the device symmetry or toward the source side only to selectively reduce the source-side underlap barrier, that is, in the form of AymD-kS architecture. Due to the limited usage of high-permittivity material only on one side, the AsymD-kS architecture further helps in reducing the overall trap charges, Miller capacitance (C_{GD}), and the high-k mobility degradation in comparison to the symmetric device (SymD-k). However, in certain applications such as pass transistor logic and static random-access memory (SRAM), where biasing changes dynamically; the bidirectional current characteristics of AsymD-kS architecture need equal attention.

A three-dimensional and top (2D) view of an optimized symmetric and asymmetric dual-k spacer tri-gate structures are shown in Figure 3.2a and b

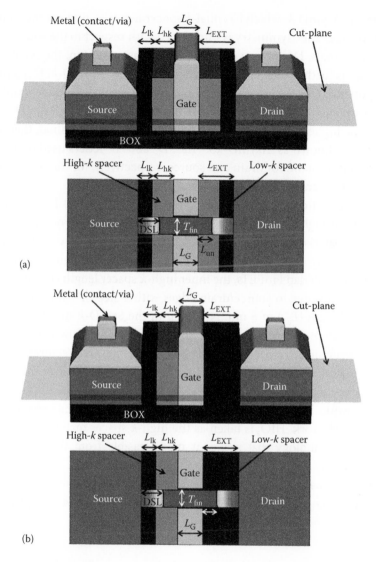

FIGURE 3.2   3D and top (2D) view of proposed (a) symmetric, SymD-*k* and (b) asymmetric dual-*k* spacer (AsymD-*k*S) tri-gate FinFETs.

respectively. These structures are in contrast to the conventional one; where a single/low-*k* spacer material is used all over the extension region, that is, from the gate edge to S/D edges. Because of the coexistence of both high-*k* and low-*k* spacers in the extension region, the architecture is abbreviated as dual-*k* architecture. It consists of an inner high-permittivity and an outer low-permittivity spacer material. Similarly, the asymmetric structure is termed as AsymD-k, which has dual-k spacer material either at the source or drain side. The asymmetry is introduced with respect to the source and drain terminals. Depending on the applied bias condition, the AsymD-k structure is further categorized as "AsymD-kS" and "AsymD-kD." If a positive potential (for the n-type FinFET) is applied at the dual-k side terminal with respect to the low-k side terminal, the structure is called "AsymD-kD;" and if the higher potential is applied at the low-k side terminal, then the structure is known as "AsymD-kS." The "D" and "S" letters used in device nomenclature stand for drain and source, respectively, for the reason that the higher potential terminal (in n-type) acts as a drain.

In a conventional device, silicon dioxide (SiO$_2$, $k = 3.9$) is adopted as a spacer throughout the G-S/D extension region. On the other hand, if a high-k material replaces the SiO$_2$ in the conventional device, then it is called a "purely high-k" device. To analyze the effect of high-permittivity spacers on underlap FinFETs, the inner high-k spacer length (L_{hk}) is varied from the gate edge to source/drain edges with a step of 4 nm for a fixed underlap length (L_{un}) of 8 nm. Note that in the SymD-k structure, L_{hk} of 0 nm and 20 nm correspond to the conventional single/low-k and purely high-k structures, respectively. All the considered architectures in this research work are summarized in Table 3.1. For high-k spacer materials, Si$_3$N$_4$ ($k = 7.5$), HfO$_2$ ($k = 22$), and TiO$_2$ ($k = 40$) are considered throughout the work. Except for the spacer permittivity variations presented in this research work, HfO$_2$ is taken as the high-k spacer material due to its advanced technology that makes it more feasible from the fabrication point of view.

The physical and electrical parameters are calibrated to meet the specifications according to the ITRS projections for 14 nm physical

TABLE 3.1 Nomenclature for Various Device Architectures and Their Characterization

Architectures	Characterization
Conventional	Single/low-k spacer throughout G-S/D extension region
SymD-k	Symmetric dual-k spacer
AsymD-k	Asymmetric dual-k spacer either at source or drain side
AsymD-kS	Dual-k spacer at source-side only
AsymD-kD	Dual-k spacer at drain-side only
High-k	Single/high-k spacer throughout G-S/D extension region

TABLE 3.2   ITRS Projections (2012) for High Performance
Devices in Year 2017 (ITRS)

Device Parameters	Abbreviations	ITRS Projections Value
Physical gate length	L_G	14 nm
Eq. oxide thickness	EOT	0.72 nm
Fin thickness	T_{fin}	9.4 nm
Fin height	H_{fin}	20 nm
Supply voltage	V_{DD}	0.75 V
Channel doping	N_A	$1 \times 10^{16}\,cm^{-3}$
Source/Drain doping	N_D	$1 \times 10^{20}\,cm^{-3}$
Threshold voltage	V_{th}	~230 mV

gate length (L_G) as summarized in Table 3.2 (ITRS). Accordingly, the fin thickness (T_{fin}), fin height (H_{fin}), and equivalent oxide thickness (EOT) are adopted as 9.4 nm, 20 nm, and 0.72 nm, respectively. The metal-gate work functions are tuned to 4.45 eV for *n*-type and 4.77 eV for *p*-type to achieve a requisite threshold (V_{th}) of ~230 mV at the supply voltage of 750 mV. The S/D extension region uses Gaussian-doping profiles followed by a lateral doping gradient of 3 nm/decade, such that the dopant-segregation length (DSL) is 12 nm. The S/D extension length (L_{EXT}) is taken as 20 nm (i.e., greater than the physical gate length). The channel and underlap region are lightly doped with a concentration of $1 \times 10^{16}\,cm^{-3}$ to reduce random dopant fluctuations (RDFs) (Colinge 2008). The raised source/drain regions have been formed to reduce the parasitic resistance associated with thinner fins. Moreover, to consider the gate-to-source/drain capacitance, metal contacts are taken. The gate-electrode thickness (T_G) is nearly twice the L_G value (Yang et al. 2007). The inner high-*k* spacer (L_{hk}) and outer low-*k* spacer length (L_{lk}) are tuned to 12 and 8 nm, respectively, for L_{un} of 8 nm. The thickness of the buried-oxide (BOX) layer is taken as 50 nm.

3.3 TCAD SIMULATION METHODOLOGY

For 3D device realization and extensive mixed-mode circuit simulations, Synopsys TCAD is used (TCAD). To include the quantum confinement of carriers in a thin silicon channel, the quantum potential model is adopted instead of only drift diffusion. The direct tunneling model is included that considers all the gate leakages including the edge-gate direct tunneling current. The Philips unified mobility model is enabled that accounts for both impurity and carrier–carrier scattering mechanisms. Besides this, the high-k Lombardi mobility model has been activated to account for high-k mobility degradation at the Si/spacer interface at a high transverse electric field. In ultrascaled devices at a high electric field, the carrier drift velocity is no longer proportional to the electric field; instead, the velocity saturates to a finite speed, v_{sat}. Therefore, the *Canali* model is enabled to account for the same.

3.4 INNER HIGH-k SPACER LENGTH OPTIMIZATION

For an underlap structure, introduction of high permittivity (k) spacer material modulates the charge dynamics in the underlap regions. For an abrupt junction, the inner high-k spacer extends up to the junction (Nandi et al. 2012), but such a profile is unrealistic from the fabrication point of view. Moreover, the variability issues with low straggle have also become prominent. For Gaussian lateral extension of S/D, the device performance in tri-gate FinFETs largely depends on the underlap length and lateral doping straggle (σ_L) (Zhao 2008). If high-permittivity materials are used as spacers to enhance the device performance, then its permittivity and length should be carefully optimized.

3.4.1 Charge Modulation by Spacer Engineering

This section demonstrates the optimization strategy of inner high-k spacer length to maximize the proposed device performance. Figure 3.3a and b plots the drive ($V_{GS} = V_{DS} = V_{DD}$) and subthreshold leakage current ($V_{GS} = 0$, $V_{DS} = V_{DD}$) in SymD-k and AsymD-kS architectures, respectively. These plots are considered as a function of high-k spacer permittivity and the length varied from gate edge ($L_{hk} = 0$ nm) to source and/or drain edges ($L_{hk} = L_{EXT} = 20$ nm) for a fixed L_{un} of 8 nm. It is observed that in both the SymD-k and AsymD-kS architecture, the I_{ON} increases with an increase in L_{hk} until an optimal point of 12 nm is reached. Beyond this point, I_{ON} starts degrading. The behavior can be easily understood by the two-dimensional electron-density variations in silicon film as a function of the inner high-k spacer length as shown in Figure 3.4. In this, L_{hk} is also varied from the gate edge to the S/D edge with a step of 4 nm.

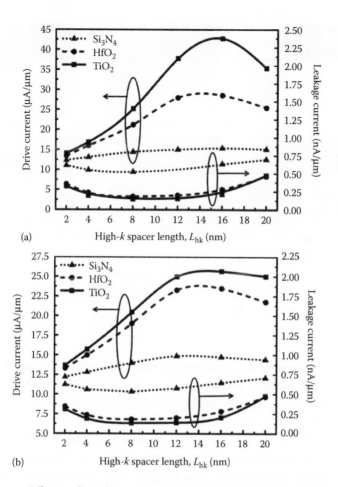

FIGURE 3.3   Effect on drive ($V_{GS} = V_{DS} = V_{DD}$) and sub-threshold leakage current ($V_{GS} = 0$, $V_{DS} = V_{DD}$) in (a) SymD-*k* and (b) AsymD-*k*S structures as a function of L_{hk} with different high-*k* permittivity materials.

It is observed from the conventional device (Figure 3.4a, $L_{hk} = 0$ nm) that the underlap and its nearby laterally diffused regions have low carrier concentration; hence, an energy barrier is formed. This barrier restricts the carrier flow from source to drain that necessitates the usage of high-*k* spacer material. The energy barrier is reduced by gate-induced fringe field lines that enhance the drive current. Figure 3.4b–d depict that the carrier concentration increase beneath the spacer interface with an increase in L_{hk} till this

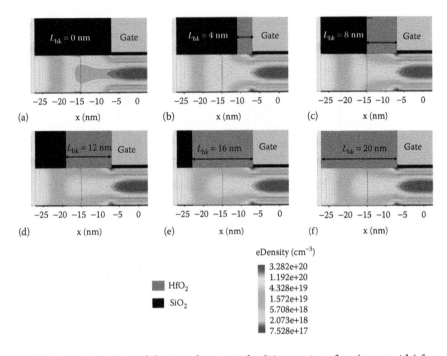

FIGURE 3.4 Variation of electron density at the Si/spacer interface (source side) for $V_{GS} = V_{DD}$ and $V_{DS} = 0$ V as a function of L_{hk} with (a) $L_{hk} = 0$ nm, (b) $L_{hk} = 4$ nm, (c) $L_{hk} = 8$ nm, (d) $L_{hk} = 12$ nm, (e) $L_{hk} = 16$ nm, and (f) $L_{hk} = 20$ nm.

length reaches a value of 12 nm. The increase in I_{ON} (shown in Figure 3.3) is due to this modulation of carrier concentration in the underlap and its nearby lightly doped region. The fringing field lines originate from the gate and travel through the inner high-k spacer. Due to the difference in spacer permittivity (inner/outer spacer k values), the field lines converge at the interface and terminate underneath the high-k/low-k spacer interface region. Hence, more charges accumulate under the inner/outer spacer interface that lead to a reduction in the underlap barrier, thereby resulting in higher I_{ON}. As the inner spacer permittivity increases, more convergence of field lines occurs; and thereby, increases the charge density under the dual spacer interface. Therefore, the higher field beneath the interface results in higher carrier concentration; and hence, a higher ON current.

However, beyond an optimum value of $L_{hk} = 12$ nm, the I_{ON} starts decreasing due to spreading of fringing field lines. These fringing field lines now fall on the laterally diffused S/D region that already has very high carrier concentration. Hence, the intensity of fringing field lines on underlap and its nearby lightly doped region is comparatively reduced.

Therefore, the carrier density slightly reduces, which results in lower I_{ON} in comparison to the drive current at the optimum point. The measured electron-density values ($\times 10^{18}$ cm^{-3}) at a high-/low-*k* interface (at $x = \pm 19$ nm) are 3.24, 15.58, and 11.2 for L_{hk} of 0 (Figure 3.4a), 12 (Figure 3.4d), and 20 nm (Figure 3.4f), respectively. Therefore, placement of high-*k* spacers beyond the highly doped laterally diffused area is not required, which may otherwise lead to higher trap charges and unnecessarily higher fringe capacitances. The difference in spacer permittivity at the interface region changes the electric field path that results in superior electrostatics in the OFF-state as well.

Figure 3.5a and b plots the I_{ON}/I_{OFF} current ratio variations in SymD-*k* and AsmD-*k*S architectures, respectively, as a function of L_{hk} for different

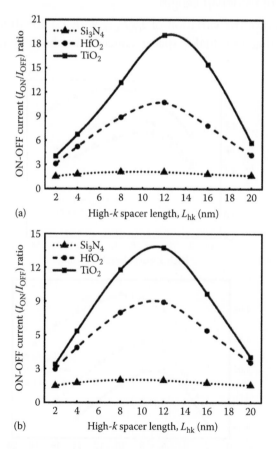

(a)

(b)

FIGURE 3.5 Effect on I_{ON}/I_{OFF} ratio in (a) SymD-*k*, and (b) AsymD-*k*S device structures normalized with respect to conventional as a function of L_{hk} with different high-*k* permittivity materials.

spacer permittivity materials. Similar to the ON and OFF characteristics, the I_{ON}/I_{OFF} current ratio increases sharply with the increase in L_{hk} and inner spacer permittivity; and beyond the optimal point of $L_{hk} = 12$ nm, it starts decreasing. The I_{ON}/I_{OFF} trends of proposed dual-k architectures are mainly due to the I_{ON} characteristics. Therefore, to obtain the best performance, the dimensions of inner high-k spacer (L_{hk}) and outer low-k spacer length (L_{lk}) are optimized to 12 nm and 8 nm, respectively. In other words, for the fixed values of σ_L and L_{un}, a maximum I_{ON}/I_{OFF} can be achieved with an extended high-k spacer length of 4 nm beyond the underlap length. It is also observed that both SymD-k and AsymD-kS structures outperform the purely high-k structure (i.e., SymD-k device when $L_{hk} = 20$ nm).

3.4.2 Effect of Underlap Length

Figure 3.6 shows the variation of I_{ON}/I_{OFF} with L_{hk} for different L_{un}. It is observed that the dual-k spacer concept (for both symmetric and

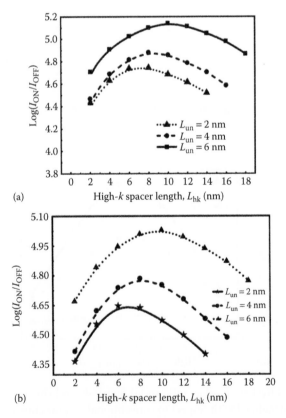

FIGURE 3.6   (a, b) Shifting of maximum I_{ON}/I_{OFF} point with different underlap values.

asymmetric) and its optimization strategy holds good even for smaller underlap lengths. It shows the I_{ON}/I_{OFF} ratio trends for $L_{un} = 2, 4,$ and 6 nm keeping all the other parameters constant. It is observed from the obtained trends that for different underlap lengths, only the inner high-*k* spacer length has to be changed. For the optimal point in L_{un} of 2, 4, and 6 nm, the L_{hk} should be 6, 8, and 10 nm, respectively; which suggests L_{hk} to be 4 nm more than the underlap length. As the underlap length increases, the I_{ON}/I_{OFF} ratio will increase with an improved DIBL and subthreshold swing (*SS*).

3.5 FABRICATION METHODOLOGY OF SymD-*k* AND AsymD-*k*S FinFETs

This section briefly provides the proposed methodology of fabricating dual-*k* spacer architectures under study. The key process steps involved in fabricating the proposed SymD-*k* and AsymD-*k*S device architectures are the dual-*k* spacer technology for creating an underlap region at S/D as depicted in Anderson et al. (2006) and the high-*k* dielectric spacer patterning as used in Xiong et al. (2004). Both the SymD-*k* and AsymD-*k*S tri-gate FinFET fabrication flow starts with either bulk or SOI wafers for fin formation with required thickness and height. Thereafter, channel doping is performed by using a masked ion implantation and a sacrificial oxidation is employed prior to gate oxidation to eliminate etch damages. Next, the gate dielectric is grown and metal gate is deposited with requisite work function. It is appropriate to tune the threshold voltage of the structure by using a gate material that has proper work function. After gate formation, high-*k* (preferably HfO_2) offset spacers are formed on both sides, to achieve symmetric doping profiles and underlap lengths. Thereafter, S/D extension regions are formed after the high-*k* offset spacer using low-energy implantation. Subsequently, in the SymD-*k* structure, the raised S/D region is grown by using selective epitaxial growth. It helps to reduce the parasitic resistance associated with thin fins. The thermal anneals are used for dopant activation. At last, the silicidation and metallization are carried out.

Because of symmetric nature, the SymD-*k* architecture is quite simpler to fabricate than its asymmetric counterpart. Asymmetric dual-*k* spacer tri-gate (AsymD-*k*S) FinFET architecture consists of an optimized inner high-*k* and an outer low-*k* spacer material only at the source side, as shown in Figure 3.7. Therefore, the major fabrication challenge associated with the AsymD-*k*S device is to form an asymmetric dual-*k* spacer along with the symmetrical underlap regions on both sides. Several researchers over the past years have

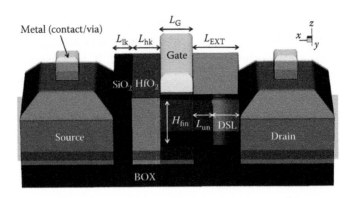

FIGURE 3.7   3D view of asymmetric dual-*k* spacer at source (AsymD-*k*S) tri-gate FinFET.

proposed the method used for fabricating single and/or double dielectric asymmetric spacers (Cheng et al. 2011), but it is intentionally used to form unequal source/drain (S/D) underlap regions. Symmetrical S/D underlap regions can only be formed by using symmetrical spacers that provide an offset of dopant implant. Figure 3.8 illustrates the proposed process flow of introducing two spacers only on one of the source or drain side with equal underlap regions on both sides.

After the gate formation, high-*k* spacers are formed on both sides that provide the offsets. S/D extensions are formed using tilt angle implants to achieve symmetric doping profiles and underlap values. Thereafter, a photo-resist layer is formed over the structure. This photo-resist mask and the gate structure shield the source-side inner high-*k* spacer from damage by the angled ion implant (Lee and Mocuta 2004). The angled

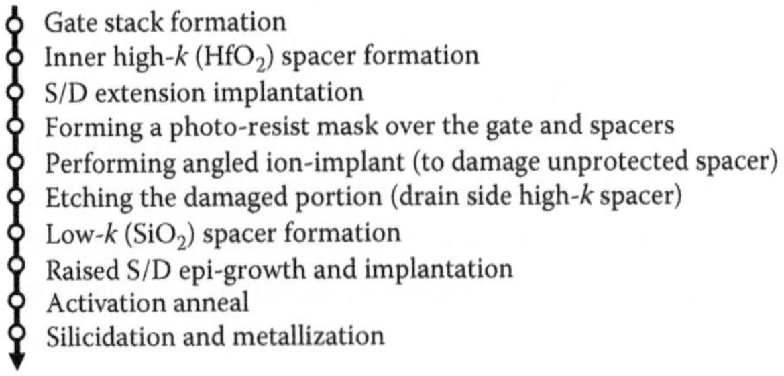

- Gate stack formation
- Inner high-*k* (HfO$_2$) spacer formation
- S/D extension implantation
- Forming a photo-resist mask over the gate and spacers
- Performing angled ion-implant (to damage unprotected spacer)
- Etching the damaged portion (drain side high-*k* spacer)
- Low-*k* (SiO$_2$) spacer formation
- Raised S/D epi-growth and implantation
- Activation anneal
- Silicidation and metallization

FIGURE 3.8   The proposed fabrication flow for AsymD-*k*S tri-gate FinFET structure.

ion implant is the key technology to selectively damage the unprotected high-k spacer so as to subsequently fabricate the inner high-k spacer at the source side (Wei et al. 2008). After this, the damaged spacer (i.e., drain-side high-k spacer) is subsequently removed by reactive ion etching (RIE). Now, we have a source-side high-k spacer with equal underlap lengths. The low-k (SiO_2) spacers are formed over the remaining portions; and then, the raised S/D region is grown by selective epitaxial growth. Thereafter, the annealing, silicidation, and metallization steps are performed.

3.6 ELECTROSTATICS AND MERITS OF DUAL-k SPACER FinFETs

For an underlap structure, the introduction of optimum high permittivity (k) spacer material modulates the charge dynamics in the underlap and channel as well. This section demonstrates the unique properties, merits, and electrostatics associated with the proposed symmetric (SymD-k) and asymmetric dual-k (AsymD-kS) FinFET structures. The proposed dual-k device architectures have an interface region (high-k/low-k) that makes it different from the conventional single/low-k and purely high-k structures. This interface region mainly improves the electrostatics of the proposed architectures as discussed in Section 3.3. The difference in spacer permittivity at the interface region changes the electric field path, which results in superior electrostatics in both ON- and OFF-state. To better understand both the symmetric and asymmetric dual-k architectures, it is necessary to first analyze them separately. Thereafter, both the proposed SymD-k and AsymD-k architectures are compared to distinguish the competing effects of high-permittivity spacers among them.

3.6.1 Symmetric Dual-k (SymD-k) Tri-Gate Architecture

The optimized symmetric dual-k spacer architecture, abbreviated as "SymD-k," consists of an inner high-k and an outer low-k spacers on both source and drain sides. The TCAD drawn schematic of the same is shown in Figure 3.9. It sharply differs from the conventional and high-k structures, where single low-k and high-k spacer materials, respectively, are used throughout the extension region.

This subsection demonstrates the brief electrostatics and merits associated with the SymD-k FinFET over the conventional and purely high-k

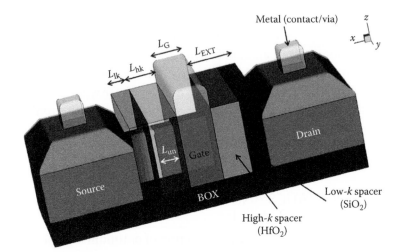

FIGURE 3.9   3D-view of symmetric dual-k spacer (SymD-k) tri-gate FinFET.

structures considering HfO_2 as the inner high-k spacer material. In comparison to the conventional and high-k spacer FinFETs, the SymD-k FinFET structure displays a higher conduction band energy under the gate region at $V_{GS} = 0$, as observed in Figure 3.10. It is because of the subdued influence of the drain electric field on the channel that substantially reduces the subthreshold leakage current.

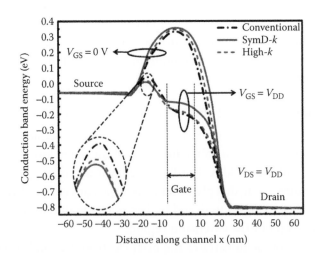

FIGURE 3.10   Conduction band profile along the channel for conventional, high-k, and SymD-k n-FinFET structures at $V_{GS} = 0$ and V_{DD} when $V_{DS} = V_{DD}$.

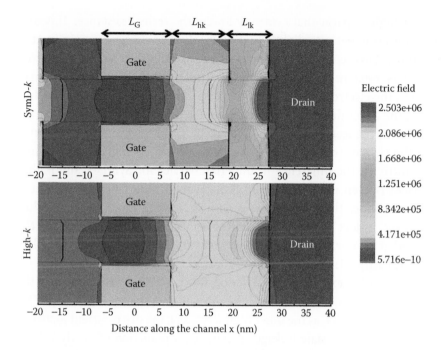

FIGURE 3.11 The electric-field contours (*xy*-plane cutting at $H_{fin} = 10$ nm) in OFF-state that demonstrates the influence of drain-field on underlap and channel region in SymD-*k* (top) and high-*k* devices (bottom).

To better understand the superior OFF-state electrostatics, Figure 3.11 presents the electric field contours at $V_{DS} = V_{DD}$, $V_{GS} = 0$ V in SymD-*k* and high-*k* structures. It clearly demonstrates that when the SymD-*k* device is in the OFF-state ($V_{DS} = V_{DD}$, $V_{GS} = 0$ V), most of the electric field lines terminate at/near the interface (high-/low-*k* spacer) region; and thereafter, the field intensity reduces as one move toward the channel region. On the other hand, in a purely high-*k* device, the drainfield intensity in the underlap and channel region is larger than that in the proposed SymD-*k* device; hence, it influences the channel region to a greater extent. Furthermore, in the channel region of SymD-*k*, it is observed that the influence of the drain field reduces even more with the further increase in inner spacer permittivity. This results in higher conduction band energy (CBE) in the OFF-state for the SymD-*k* device than that for the high-*k* device.

Moreover, in the ON-state ($V_{GS} = V_{DS} = V_{DD}$), the increased fringing field produces an accumulation of carriers in the underlap region through the inner high-k spacer that evidently lowers the series resistance. This barrier lowering permits slightly higher injection of charge carriers into the channel. Furthermore, the barrier directly under the channel is lowered to a lesser extent by the drain bias in SymD-k than in the conventional; hence, the electrostatic integrity increases, which reduces SCEs. In addition, the lower drain field influence on the underlap and channel region results in reduced DIBL. The influence of drain field reduction depends on the type of inner high-k spacer used. With the higher inner spacer permittivity, the difference in underlap barrier for SymD-k and high-k devices is even more significant. Furthermore, the CBE barrier that is directly under the gate, increases with increasing inner spacer k value; however, it does not affect I_{ON} unless it is significantly higher than the underlap barrier. Once the carriers cross the G-S underlap barrier, they can be easily transported to the drain end.

Figure 3.12a examines the device performance parameters to reveal the effects of fringing fields through the inner high-k spacer. For the SymD-k FinFET (with HfO$_2$ as the inner high-k spacer), the drive current increases 2.4× with an OFF-state leakage current (I_{OFF}) reduction of nearly 77% as compared with the conventional one. Moreover, the SymD-k structure shows 25% improvement in $\log_{10}(I_{ON}/I_{OFF})$, with reduced DIBL (~55%) and subthreshold swing (~5.8). It is also observed from Figure 3.12a that the proposed SymD-k device outperforms the purely high-k spacer device in terms of all the performance metrics. The output characteristics of the proposed device compared with the conventional structure are plotted in Figure 3.12b. The results depict that the drain current for a given V_{GS} in the saturation region is almost constant for the SymD-k device compared to the conventional one; and hence, larger output impedance (or, lower output conductance) is achieved.

3.6.2 Asymmetric Dual-k (AsymD-k) Tri-Gate Architecture

In asymmetric dual-k spacer (AsymD-k) architecture, the optimal inner high-k spacer material is used only at the source side targeting to reduce the G-S underlap barrier. The asymmetric characteristics are measured by applying the bias on both the terminals as discussed earlier in Section 3.2. To implement CMOS logic gates, the asymmetry with respect to S/D does not affect its functionality, although the power and delay metrics might get affected. Asymmetry introduced with respect to S/D terminals could be beneficial in circuits that use pass transistors such as 6T SRAM cell

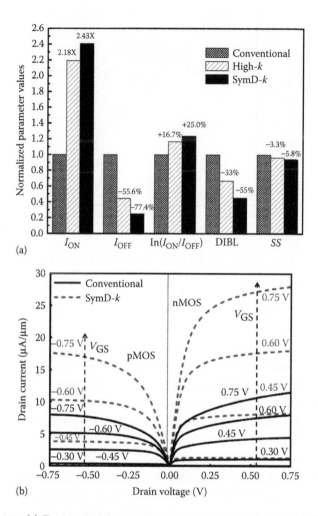

FIGURE 3.12 (a) Design metrics comparison among conventional, high-*k*, and SymD-*k* *n*-FinFETs. (b) I_D-V_{DS} comparison between conventional and SymD-*k* FinFET structure with increasing V_{GS}.

that faces a trade-off between read and write stabilities. Therefore, the proposed asymmetric dual-*k* spacer architecture helps in mitigating the read/write conflict. This subsection describes the unique electrostatics and properties associated with the AsymD-*k* structure. Various advantages of the AsymD-*k*S over the conventional single-*k* spacer FinFET are also discussed.

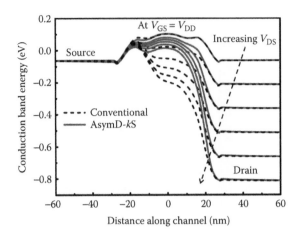

FIGURE 3.13   CBE profile of the conventional and AsymD-kS devices with increasing V_{DS} at $V_{GS} = V_{DD}$.

Figure 3.13 shows the conduction band energy with increasing drain to source potential (V_{DS}) of conventional and AsymD-kS FinFET structures at $V_{GS} = V_{DD}$. It is observed that the CBE barrier directly under the gate is lowered to a lesser extent by the drain bias in AsymD-kS than conventional; hence, the electrostatic integrity increases. In the AsymD-kS structure, the increased source coupling competes with the gate to keep the channel potential near zero volt, illustrating how source-side injection governs the drain current in a short device and reduces the effect of drain potential directly under the channel (Kencke et al. 2000). In other words, the gate voltage accumulates electrons (holes) in the n-type (p-type) device on the source side through the inner high-k spacer. This evidently lowers the source series resistance and accounts for the injection of carriers into the channel. The reduced influence of the drain field depends on the type of inner high-k spacer used at the source side. The barrier directly under the gate increases with increasing source-side inner spacer k value while maintaining the low-k spacer on the drain side.

Figure 3.14 shows the variation in the CBE profile along the channel for conventional, AsymD-kS, and AsymD-kD n-FinFET structures at $V_{GS} = 0$ and V_{DD}, when $V_{DS} = V_{DD}$. It is clear from the CBE profile that the AsymD-kS structure displays the highest conduction band energy under the gate region at $V_{GS} = 0$ that substantially reduces the subthreshold leakage. When potential is applied to the gate terminal, the conduction band energy of the channel near the source (under the G-S interface) reduces more in AsymD-kS as compared to the other two structures; hence, ON-current increases. Less barrier lowering is observed throughout the channel region,

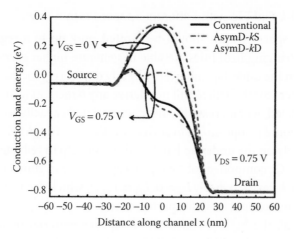

FIGURE 3.14 Variation of CBE along the channel for conventional, AsymD-*k*S, and AsymD-*k*D *n*-FinFET structures at $V_{GS} = 0$ and V_{DD} when $V_{DS} = V_{DD}$.

which results in better electrostatic control of the gate over the channel and reduced SCEs.

The AsymD-*k*D FinFET also presents marginally better results than the conventional one. The output characteristics of the proposed AsymD-*k*S device compared with the conventional structure are shown in Figure 3.15 at different V_{GS} ranging from 0 V to 0.75 V (V_{DD}) with a step of 0.15 V.

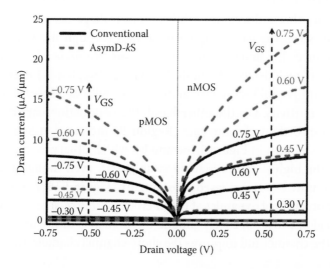

FIGURE 3.15 I_{DS}-V_{DS} comparison between conventional and AsymD-*k*S FinFET structure with increasing V_{GS}.

3.6.3 Dual-k FinFETs with Different Spacer Permittivity

The high-k spacer length should be limited only up to a region where the strong fringing field lines are able to generate higher carrier density, especially, in the underlap and nearby lightly doped region. Increasing the high-k spacer length further will not derive any benefit in terms of carrier density; instead, will increase the fringe-associated capacitances (C_{if}, C_{of}). Thus, L_{hk} should be optimized in such a way that the fringe field lines concentrates only on the underlap barrier. This can help in reducing both the $R_{S/D}$ and the C_{fr} component of total gate capacitance in comparison to the purely high-k device (where, $L_{lk} = 0$, $L_{hk} = L_{EXT}$), which can result in better device and in turn circuit delay performance.

In general, the high-permittivity (k) spacer material modulates the underlap barrier (Sachid et al. 2008); while in the proposed dual-k spacer, the fringing field modulates not only the underlap barrier, but, it also affects the charge dynamics within the channel. This section demonstrates the brief electrostatics, merits, and the current characteristics associated with the dual-k spacer FinFET architectures with an increase in inner spacer permittivity. In comparison to the conventional one, Figure 3.16 shows the conduction band energy profile along the channel of SymD-k, AsymD-kS, and purely high-k device architectures with different inner high-k spacer materials at the xy-plane cutting $H_{fin} = 10$ nm. As expected, in the OFF-state (when $V_{GS} = 0$ V, $V_{DS} = V_{DD}$), the conduction band energy increases with an increase in inner spacer permittivity that reduces the subthreshold leakage current. Moreover, in the ON-state, the gate fringing field lines through the inner high-k spacer reduces the underlap barrier that leads to a higher I_{ON}. Similar to the purely high-k device, the underlap barrier (Region-A) decreases in SymD-k and AsymD-kS device architectures with an increase in inner spacer permittivity. On the other hand, the conduction barrier (in the ON-state) directly under the gate (Region-B) increases in SymD-k and AsymD-kS (shown in Figure 3.16a and b); which otherwise remains the same in the high-k device (Sachid et al. 2008), even though the inner spacer permittivity is increased substantially (Figure 3.16c). It is because of the favorable field dynamics at the interface of source-side high-k/low-k spacer regions, that effectively subdue the influence of the drain electric field on the channel and improves the short-channel characteristics.

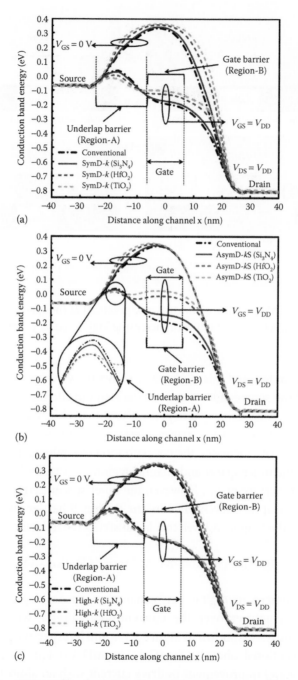

FIGURE 3.16 CBE profile along the channel (*xy*-plane cut at H_{fin} = 10 nm) for (a) SymD-*k*, (b) AsymD-*k*S, and (c) high-*k* FinFET structures in ON and OFF-state with increasing spacer permittivity.

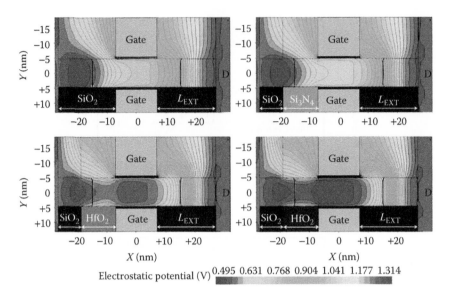

FIGURE 3.17 The variation of electrostatic potential at the spacer interface and inside the channel (in ON-state) with increasing high-k spacer permittivity.

To better understand this, Figure 3.17 shows the variation of electrostatic potential at the source-side spacer interface and inside the channel (in the ON-state) with increasing spacer permittivity. It is observed that the equipotential lines deviate at the spacer interface in the ON-state. As the inner spacer permittivity increases, the concentration of equipotential lines at the interface of the inner spacer and channel region increases; which ultimately results in an increase of I_{ON}. However, the inner spacer permittivity can be increased only up to a level; wherein, the Region-B barrier does not attain a level above the Region-A level, that is, satisfying the following condition:

$$\text{Barrier (Region-B)} \leq \text{Barrier (Region-A)}$$

The improvement in digital performance metrics using SymD-k and AsymD-kS structures over the conventional FinFET with increasing spacer permittivity is shown in Figure 3.18a and b respectively. Compared to the conventional one, the SymD-k (AsymD-kS) FinFET shows 1.3–3.3 × (1.3–2.2×) improvement in drive current with an almost 40%–80% (35%–84%) reduction in the OFF-state leakage current (I_{OFF}) when the inner spacer permittivity (k) is varied from 7.5 to 40. Moreover, the SymD-k (AsymD-kS) structure with TiO$_2$ ($k = 40$), shows up to 29.6% (27.7%)

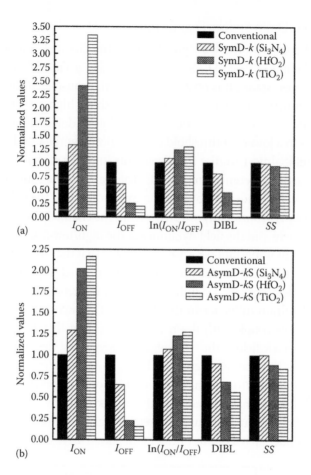

FIGURE 3.18   (a) SymD-*k* and (b) AsymD-*k*S device performance improvement in comparison to conventional FinFET with different inner spacer permittivity.

improvement in $\log_{10}(I_{ON}/I_{OFF})$, with reduced DIBL and subthreshold slope of 69% (43.5%) and 7.5% (15.4%), respectively. Therefore, SymD-*k* and AsymD-*k*S devices with higher inner spacer permittivity comprehensively outperform the conventional, high-*k*, and dual-*k* devices with lower inner spacer permittivity.

3.7 A COMPARATIVE ANALYSIS BETWEEN SYMMETRIC AND ASYMMETRIC SPACER ARCHITECTURES

This section provides an in-depth analysis of the various competing effects of a high-*k* spacer (HfO_2) implemented on the source and/or drain side.

3.7.1 Source/Drain Spacer Engineering

Figure 3.19 compares the CBE profile along the channel among the conventional, SymD-k, AsymD-kS, and AsymD-kD n-FinFET structures in the ON- and OFF-states. It is observed that all the considered dual-k spacer devices (symmetric as well as asymmetric) exhibit higher CBE barrier in the OFF-state over the conventional single/low-k device that encouragingly results in a lower subthreshold current. Both SymD-k and AsymD-kS devices exhibit almost similar improvements in terms of OFF-state electrostatics. The major differences in the CBE profile among the considered structures are observed under the gate region in the ON-state. The CBE profile or charge density is prominently affected within the channel. This is due to the spacer placement and the field dynamics at the high/low-k spacer interface.

From this, an important observation is made that the CBE barrier (in the ON-state) of AsymD-kS directly under the gate is much higher in comparison to other devices, which signifies a subdued influence of the drain field. Similarly, the drain influence in the AsymD-kD structure increases as the gate fringe lines couple with drain fringe lines; and therefore, strengthen the vertical electric field at the G-D edge. Consequently, the drain field completely controls the channel; thereby, reducing the gate electrostatic control in comparison to the conventional device that is not often preferable by device engineers. The combinational effect of the two asymmetric structures can be observed in SymD-k architecture; wherein, the gate control increases, which reduces the underlap barrier resulting in increased drive current.

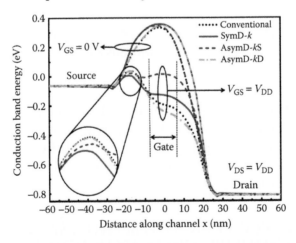

FIGURE 3.19 Conduction band profile along the channel for conventional, SymD-k, AsymD-kS, and AsymD-kD n-FinFET structures at $V_{GS} = 0$ and V_{DD} when $V_{DS} = V_{DD}$.

3.7.2 Current Characteristics

This section is devoted towards the performance evaluation of the FinFET devices using parameters such as drive current, gate leakage current, subthreshold current, and short-channel parameters. Furthermore, several effects due to source- and drain-side spacers and their combined effect on the symmetric device are also investigated. Figure 3.20 plots the drive current and subthreshold current component as a function of the high-*k* spacer length varied from the gate edge ($L_{hk} = 0$ nm) to the source and/or drain edges ($L_{hk} = L_{EXT} = 20$ nm). The drive current in both the SymD-*k* and AsymD-*k*S devices increases with the increase in L_{hk} up to an optimal point of 12 nm, beyond which, it starts decreasing. However, in the AsymD-*k*D device structure, it is throughout constant, when L_{hk} is extended from the gate edge to the drain edge. It is also observed from Figure 3.20 that the AsymD-*k*D device leads to a higher subthreshold leakage current in comparison to its counterpart AsymD-*k*S device. However, both SymD-*k* and AsymD-*k*S structures show overlapped leakage current behavior. The subthreshold leakage component in all the considered structures is less than the conventional one because of better electrostatic integrity. It is observed from the obtained results that the source-side spacer mainly governs the charge transport from source to drain; however, the drain-side spacer helps to enhance the current magnitude in SymD-*k* compared to AsymD-*k*S.

Apart from the subthreshold leakage current (I_{SUB}), the gate tunneling current (I_G) is considered to be a dominant leakage current component

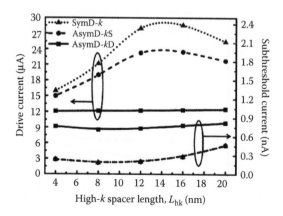

FIGURE 3.20   Effect on drive current and subthreshold current in SymD-*k*, AsymD-*k*S, and AsymD-*k*D device structures as function of L_{hk} considering HfO$_2$ as inner spacer permittivity materials.

that may affect the device performance. The gate leakage current is further composed of the gate-to-channel tunneling current (I_{GC}) and the edge direct tunneling current (I_{EDT}). When the device operates in the ON-state ($V_{GS} = V_{DD}$), I_{GC} dominates due to the direct tunneling of carriers from the channel to the gate. However, I_{EDT} appears through the gate and drain/source extension region, both in the ON-state [I_{EDT-ON} at ($V_{GS} = V_{DD}$, $V_{DS} = 0$)] and OFF-state [$I_{EDT-OFF}$ at ($V_{GS} = 0$, $V_{DS} = V_{DD}$)]. The $I_{EDT-OFF}$ component is much smaller than I_{EDT-ON}. Therefore, the total gate-tunneling current (I_{G-ON}) in the ON-state is the sum of I_{GC} and $2 \times I_{EDT-ON}$ (Bansal et al. 2004).

Figure 3.21a and b show the variations of ON- and OFF-state gate tunneling current, respectively, with L_{hk} for the three considered dual-k

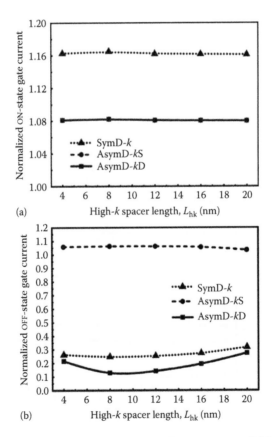

(a)

(b)

FIGURE 3.21 Effect on (a) ON-state gate tunneling current and (b) OFF-state gate tunneling current as a function of L_{hk} in SymD-k, AsymD-kS, and AsymD-kD device structures.

structures normalized with respect to the conventional one. As expected, both the AsymD-*k*S and AsymD-*k*D architectures show the same $I_{\text{G-ON}}$ with almost overlapping values at different spacer lengths. Moreover, an around 8% increase in $I_{\text{G-ON}}$ is observed in both asymmetric architectures as compared to the conventional one. Since the gate dielectric material and the thickness are the same in all the devices, the I_{GC} component is almost constant; and hence, the difference in $I_{\text{G-ON}}$ is mainly due to the increase in $I_{\text{EDT-ON}}$ due to the incorporation of high-*k* spacers. It is well known that $I_{\text{EDT-ON}}$ flows only near the gate edges; therefore, an increase in L_{hk} does not reflect any change in it. The combined effect of spacers on both sides in the SymD-*k* device results in a constant $I_{\text{G-ON}}$ that is nearly 16% higher than the conventional one. On the other side, the $I_{\text{EDT-OFF}}$ in the OFF-state ($V_{\text{GS}} = 0$, $V_{\text{DS}} = V_{\text{DD}}$) reduces in SymD-*k* and AsymD-*k*D structures; whereas, it remains the same and marginally high in the AsymD-*k*S device as compared to the conventional one.

Figure 3.22 shows the normalized $I_{\text{ON}}/I_{\text{OFF}}$ variation with L_{hk} in three considered architectures. For both SymD-*k* and AsymD-*k*S devices, the maximum $I_{\text{ON}}/I_{\text{OFF}}$ is obtained at the high-permittivity spacer length L_{hk} of 12 nm because of the superior ON- and OFF-state electrostatics. On the other hand, in the case of the AsymD-*k*D structure, it remains constant.

Figure 3.23 examines the improvement observed in the device structures at an optimized inner high-*k* spacer length of 12 nm normalized with respect to the conventional one. For SymD-*k* and AsymD-*k*S FinFETs, the drive current increases 2.43× and 2.02×, respectively, while

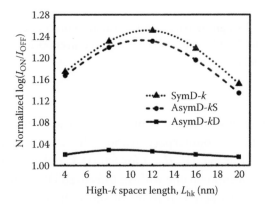

FIGURE 3.22 Variations in normalized $\log(I_{\text{ON}}/I_{\text{OFF}})$ ratio as function L_{hk} for different dual-*k* spacer devices.

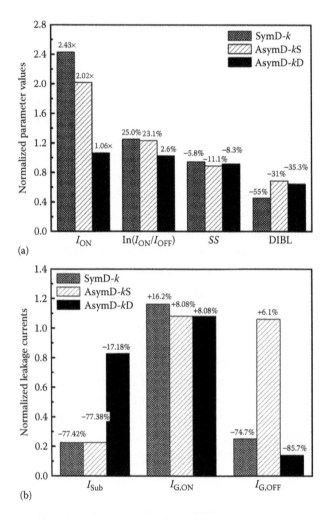

(a)

(b)

FIGURE 3.23 (a) Design metrics and (b) leakage currents comparison among SymD-*k*, AsymD-*k*S, and AsymD-*k*D device structures normalized with respect to the conventional one.

a marginal (~6%) improvement is observed in the AsymD-*k*D structure over its conventional counterpart. Almost the same increment, that is, ~25%, is observed in $\log_{10}(I_{ON}/I_{OFF})$ for SymD-*k* and AsymD-*k*S in comparison to the conventional one with an improved DIBL and SS. Although AsymD-*k*D performs better in terms of short-channel characteristics, it shows marginal improvement in ON and OFF currents as compared to the conventional structure.

3.8 SUMMARY

This chapter introduces the dual-*k* spacer concept that uses an optimal high-*k* spacer length for enhancing the device performance of an underlap FinFET. Current characteristics and SCEs metrics of the proposed symmetric and asymmetric dual-*k* spacer devices are presented and compared with conventional single/low-*k* as well as purely high-*k* FinFET structures. To evaluate the optimal high-*k* point, the effect on drive current, leakage current, and their ratio are discussed with respect to high-*k* spacer length and spacer material. Due to fringing field through the spacer in dual-*k* architectures, not only the underlap region is affected but the charge density within the channel is also modulated significantly. The favorable field dynamics and charge modulation due to the high-*k*/low-*k* spacer interface of the device is analyzed to evaluate its overall performance. It also demonstrates the electrostatics and merits associated with the proposed SymD-*k* and AsymD-*k*S FinFET structures over the conventional one with varying inner high-*k* spacer permittivity. From this, an important and novel observation is made that the CBE barrier (in the ON-state) directly under the gate increases in dual-*k* architectures; which otherwise remains the same in the high-*k* device, even though, the inner spacer permittivity is increased substantially. A detailed comparative analysis helps to examine the ON- and OFF-state electrostatics with high permittivity sidewall spacers to reveal competing effects among the symmetric and asymmetric dual-*k* spacer devices. It is observed that the source-side spacer mainly governs the charge transport from source to drain; however, the drain-side spacer helps to enhance the current magnitude in SymD-*k* compared to AsymD-*k*S. Overall, the proposed SymD-*k* and AsymD-*k*S architectures exhibit excellent device performance that would anticipate better circuit/SRAM performance.

REFERENCES

Agrawal, S. and J. G. Fossum, On the suitability of a high-*k* gate dielectric in nanoscale FinFET CMOS technology, *IEEE Trans. Electron Dev.*, 55(7), 1714–1719, 2008.

Anderson, B. A., A. Bryant, W. F. Clark, and E. J. Nowak, Low capacitance FET for operation at subthreshold voltages, U.S. Patent 70092652006, March 7, 2006.

Bansal, A., B. Paul, and K. Roy, Impact of gate underlap on gate capacitance and gate tunneling current in 16 nm DGMOS devices, in *Proceedings of IEEE SOI Conference*, IEEE, South Corolina, pp. 94–95, October 2004.

Chen, Q., L. Wang, and J. D. Meindl, Fringe-induced barrier lowering (FIBL) included threshold voltage model for double-gate MOSFETs, *Solid. State. Electron.*, 49(2), 271–274, 2005.

Cheng, K., X. Li, and R. S. Wise, Method of forming asymmetric spacers and methods of fabricating semiconductor device using asymmetric spacers, U.S. Patent 2011/0108895, A1, May 12, 2011.

Colinge, J. P., *FinFETs and other Multi-Gate Transistors*, New York: Springer-Verlag, 2008.

Fossum, J. G., M. M. Chowdhury, V. P. Trivedi, T.-J. King, Y. K. Choi, J. An, and B. Yu, Physical insights on design and modeling of nanoscale FinFETs, in *IEDM Technology Digest*, IEEE, Washington, DC, pp. 679–682, 2003.

International Technology Roadmap for Semiconductors. Available: http://public. itrs.net, 2013.

Jan, C. H. et al., A 22 nm SoC platform technology featuring 3-D trigate and high-k/metal gate, optimized for ultra low power, high performance and high density SoC applications, in *IEDM Technical Digest*, IEEE, New York, p. 44, 2012.

Kencke, D. L., Q. Ouyang, W. Chen, H. Wang, S. Mudanai, A. Tasch, and S. K. Banerjee, Tinkering with the well-tempered MOSFET: Source–channel barrier modulation with high-permittivity dielectrics, *Superlattices and Microstruct.*, 27(2–3), 207–214, 2000.

Koley, K., A. Dutta, S. K. Saha, and C. K. Sarkar, Analysis of high-κ spacer asymmetric underlap DG-MOSFET for SOC application, *IEEE Trans. Electron Dev.*, 62(3), 1–6, 2015.

Lee, B. H. and A. C. Mocuta, Method of forming asymmetric extension MOSFET using a drain side spacer, U.S. Patent 6746924, B1, June 8, 2004.

Manoj, C. R. and V. R. Rao, Impact of high-k gate dielectrics on the device and circuit performance of nanoscale FinFETs, *IEEE Electron Dev. Lett.*, 28(4), pp. 295–297, 2007.

Mohapatra, N., M. P. Desai, S. G. Narendra, and V. R. Rao, The effect of high-k gate dielectrics on deep submicrometer CMOS device and circuit performance, *IEEE Trans. Electron Dev.*, 49(5), 826–831, 2002.

Nandi, A., A. K. Saxena, and S. Dasgupta, Impact of dual-k spacer on analog performance of underlap FinFET, *Microelectronics J.*, 43(11), 883–887, 2012.

Nirmal, D., P. Vijayakumar, D. M. Thomas, B. K. Jebalin, and N. Mohankumar, Subthreshold performance of gate engineered FinFET devices and circuit with high-k dielectrics, *Microelectron. Reliab.*, 53(3), 499–504, 2013.

Rao, V. R. and N. R. Mohapatra, Device and circuit performance issues with deeply scaled high-*k* mos transistors, *J. Semicond. Technol. Sci.*, 4(1), 52–62, 2004.

Sachid, A. B., C. R. Manoj, D. K. Sharma, and V. R. Rao, Gate fringe-induced barrier lowering in underlap FinFET structures and its optimization, *IEEE Electron Dev. Lett.*, 29(1), 128–130, 2008.

Sentaurus TCAD User Manual, 2013. Synopsys, Inc. Available: www.synopsys. com, 2015.

Wei, A., G. Burbach, and D. Greenlaw, Gate structure and a transistor having asymmetric spacer elements and methods of forming the same, U.S. Patent 7354839, B2, April 8, 2008.

Xiong, Z., H. Liu, C. Zhu, and J. K. Sin, Characteristics of high-K spacer offset-gated polysilicon TFTs, *IEEE Trans. Electron Dev.*, 51(8), 1304–1308, 2004.

Yang, J. W., P. M. Zeitzoff, and H. H. Tseng, Highly manufacturable double-gate finFET with gate-source/drain underlap, *IEEE Trans. Electron Dev.*, 54(6), 1464–1470, 2007.

Zhao, H. et al., Analysis of the Effects of Fringing Electric Field on FinFET Device Performance and Structural Optimization Using 3-D Simulation, *IEEE Transaction on Electronic Devices*, 55(5), pp. 1177–1184, 2008.

CHAPTER 4

Capacitive Analysis and Dual-*k* FinFET-Based Digital Circuit Design

4.1 INTRODUCTION

Double/tri-gate FinFETs are recognized as one of the most promising successors of conventional bulk MOS devices in the sub-20 nm regime due to their excellent short-channel characteristics and reduced leakage currents (Colinge 2008). However, in addition to the advantages offered at the device level, the FinFET also offers some new challenges from circuit perspectives. The two main inherent challenges associated with the FinFET are the higher magnitude of parasitic due to its three-dimensional (3D) architecture and the fin width quantization; that limit its applicability in high-performance circuit applications due to conflicting design requirements. In conventional MOS devices, digital designers considered width, length, and area as parameters for evaluating the trade-off between transistor configuration and electrical performance. The nature of FinFET design could dramatically change the scenario.

In an underlap device, the source/drain series resistance ($R_{S/D}$) starts dominating, which limits the device drive current (I_{ON}). Incorporating high-*k* spacers can provide strong field coupling between the gate and the undoped underlap region that reduces $R_{S/D}$ (Trivedi 2005). Previous studies have shown device performance with an optimized underlap structure (Shenoy and Saraswat 2003; Schulz et al. 2004). However, they have not considered the fringe capacitance (C_{fr}) that strongly affects the switching speed in deeply scaled FinFETs. Therefore, digital circuit designers need to adapt their designs taking into account these critical issues so as to improve the overall performance in terms of device and circuit parameters such as I_{ON}, I_{OFF}, noise immunity, and the switching speed.

Most of the previous work on FinFET devices have been done at the device and process levels (Dey et al. 2008; Majumdar et al. 2011; Sharma et al. 2011). At the CAD and circuit levels, only few researchers have looked into the FinFET design issues. Several researchers have focused on the integration of high-k materials as gate dielectrics and/or spacers from the device level (Vellianitis et al. 2007; Shahrjerdi et al. 2009; Agrawal and Fossum 2010). A high-k gate dielectric could offer additional advantages such as thinner effective oxide thickness, which implies higher gate capacitance (C_{GG}) and I_{ON}. Also, the larger high-k dielectric thickness reduces the parasitic gate-source/drain (G-S/D) outer fringe capacitance (Kim et al. 2006; Manoj and Rao 2007). The fringing field phenomenon through this high-k gate dielectric has been studied by few researchers from circuit perspectives (Mohapatra et al. 2002; Agrawal and Fossum 2010). Agrawal and Fossum (2010) simulated a CMOS ring oscillator using a high-k gate dielectric and reported a modest performance enhancement for an optimized permittivity of $k \approx 20$, relative to the counterpart CMOS with a SiO_2 gate dielectric. Several research work have reported the impact of gate-sidewall fringing field in enhancing the device performance using high-k spacers. However, a comprehensive study of the 3D fringing field due to high permittivity spacers on circuit performance is still critically required. In addition, most of the reported data from circuit perspective are based on 2D device/circuit simulations. As the FinFET is a 3D device, the 2D structures remain susceptible to error. Therefore, to fully capture the effect of the 3D fringing field, the 3D tri-gate device geometry should be considered.

This chapter investigates the effect of the optimized symmetric and asymmetric dual-k structures for better logic circuit performance. This work demonstrates the suitability of high-k spacer materials for high-performance logic circuits improving the noise margin and delay simultaneously. The analysis presented in this chapter proves that the circuit delay reduces sharply with an increase in inner spacer permittivity that otherwise worsens for purely high-k structures. This chapter comprises two major sections. Section 4.2 describes the role of fringe capacitances associated with the proposed SymD-k and AsymD-kS architectures. Thereafter, the impact of capacitance on circuit performance with different inner high-k spacer materials and lengths is investigated in Section 4.3. The circuit performances are evaluated based on the static and dynamic characteristics of a CMOS inverter and a three-stage ring oscillator. Furthermore, the effect of power supply scalability on dual-k–based circuits is investigated in Section 4.4. Finally, Section 4.5 presents a brief summary of this chapter.

4.2 IMPACT OF FRINGE CAPACITANCE ON DUAL-k FinFETs

Fringe capacitance plays a dominant role in describing the circuit behavior. It is clear from the discussion presented in the preceding chapter that the performance of the conventional, high-k, and dual-k architectures is mainly dependent on the role of fringe capacitances. Incorporation of high permittivity spacers, length, and its placement modulates the charge concentration and the electrostatics of the proposed architectures in comparison to conventional and high-k architectures.

In an underlap device, total gate capacitance (C_{GG}) includes gate-to-channel (C_{GC}) and fringe capacitance (C_{fr}) components. The C_{GC} component is almost the same among the three considered structures. The introduction of high-k spacers enhances the fringe capacitance components of the structure. The total fringe capacitance consists of outer (C_{of}) and inner (C_{if}) fringe components. Since, in the strong inversion operation, C_{if} is screened by the channel; the outer fringe capacitance mainly affects the carrier modulation in the underlap and channel region in dual-k spacer architectures. In general, the dynamic circuit performance of a high-k spacer device degrades due to very high parasitic capacitances. Therefore, before going deeper into circuit analysis, this section first explores the capacitive behavior of the dual-k spacer architecture in comparison to the conventional single/low-k and purely high-k architectures.

Figure 4.1 depicts the change in C_{GG} in the proposed SymD-k, AsymD-kS, and AsymD-kD architectures normalized with respect to the conventional architecture (considering HfO$_2$ as spacer material) as a function of L_{hk}. It is observed that both SymD-k and AsymD-kS demonstrate similar capacitive behavior. As L_{hk} is increased toward the S/D edge (or toward the source edge only in the case of AsymD-kS), the outer fringe component dominates the overall gate capacitance. To better understand the role of individual capacitances and their impact on device performance, C_{GG} is further divided into gate-to-source (C_{GS}) and gate-to-drain capacitance (C_{GD}).

Figure 4.2a and b present the contribution of C_{GS} and C_{GD}, respectively, in overall C_{GG} with an increasing inner high-k spacer length. For AsymD-kS

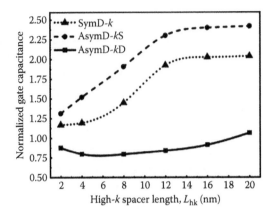

FIGURE 4.1 Effect of increasing L_{hk} on total gate capacitance (C_{GG}) in SymD-k, AsymD-kS, and AsymD-kD device structures normalized with respect to the conventional one.

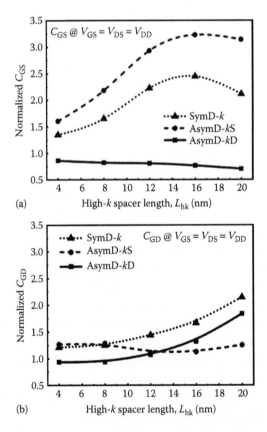

FIGURE 4.2 Effect of increasing L_{hk} on (a) C_{GS} and (b) C_{GD} in SymD-k, AsymD-kS, and AsymD-kD structures normalized with respect to the conventional one.

and SymD-*k* architectures, it is observed from Figure 4.2a that C_{GS} increases sharply up to L_{hk} = 12 nm as expected; and thereafter, decreases marginally. However, for the AsymD-*k*D architecture, although the C_{GS} component slightly decreases; it retains much lower values that are comparable to the conventional one. In the AsymD-*k*D architecture, the inner high-*k* spacer is placed only on the drain side, while the source side has the same low-*k* spacer throughout the gate edge to source edge. Therefore, if only the outer fringe component is considered, then the C_{GS} component should be constant irrespective of the change in inner high-*k* spacer length at the drain side. However, it is observed that although the overall C_{GS} component has smaller values, it continuously decreases even though the source side spacer is unchanged. This is due to the drain-side electric field that influences the channel as shown in Figure 4.3.

Note that the overall C_{GG} is divided into C_{GS} and C_{GD} from the center of the gate (at the $[x]$ = 0 nm point as shown in Figure 4.3). It is observed that with an increase in L_{hk} at the drain side, the charge density within the channel region (toward the source and drain sides) also modulates; and hence, it can be said that the drain influences the channel directly under the gate. The modulated charge directly under the gate (toward the source side) reduces the inner fringe component; and therefore, the overall C_{GS}. For shorter gate length, this effect becomes more prominent. Similarly, it is observed from Figure 4.2b that the C_{GD} component slightly fluctuates with an increase in L_{hk} of the AsymD-*k*S

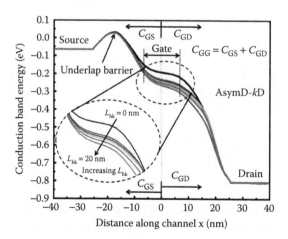

FIGURE 4.3 CBE profile along the channel for conventional and AsymD-*k*D *n*-FinFET structures with increasing L_{hk} at $V_{GS} = V_{DS} = V_{DD}$.

architecture. However, it has uniformly higher value than the conventional device because the charge modulation directly under the gate is more in the case of the AsymD-kS architecture than the AsymD-kD (shown in Figure 3.20). It is also observed from Figure 4.2b that the C_{GD} component in both the SymD-k and AsymD-kD increases steeply, which would enhance the Miller capacitance and degrade the dynamic performance of a circuit/SRAM.

To comprehensively analyze the impact of fringe capacitances with different spacer permittivities, Figure 4.4a and b show the C_{GG} variations

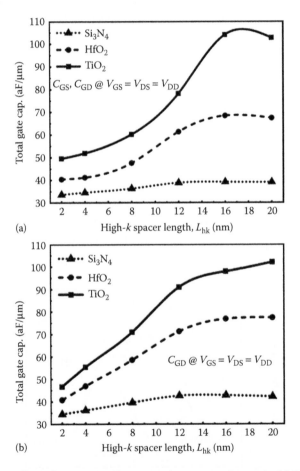

(a)

(b)

FIGURE 4.4 Effect of increasing L_{hk} on total gate capacitance in (a) SymD-k and (b) AsymD-kS structures for different k normalized with respect to the conventional one.

with L_{hk} in SymD-*k* and AsymD-*k*S considering Si_3N_4, HfO_2, and TiO_2 as inner high-*k* spacer dielectrics. In addition to the C_{GG} increment with L_{hk}, it is also observed that the fringe component excessively increases with an increase in the spacer permittivity values ranging from 7.5 to 40. Very large fringe capacitance associated with the high-*k* spacer device ($L_{hk} = 20$ nm) degrades the device performance. For TiO_2 ($k = 40$) spacer materials, the total gate capacitances in SymD-*k* and AsymD-*k*S structures are nearly 2.1× and 3.2×, respectively, that in the conventional structure with SiO_2 spacer.

Figure 4.5a and b show the variations of C_{GS} and C_{GD} in SymD-*k* and AsymD-*k*S devices, respectively, with increasing inner spacer length

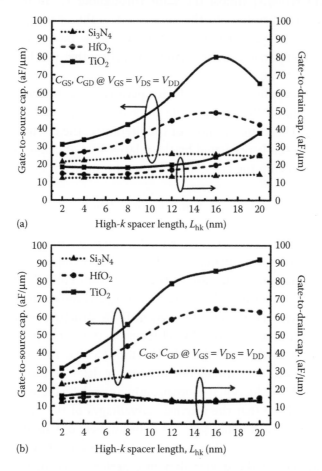

(a)

(b)

FIGURE 4.5 Effect of increasing L_{hk} on C_{GS} and C_{GD}, for different spacer permittivity materials in (a) SymD-*k* and (b) AsymD-*k*S FinFET structures.

for different spacer materials. It is observed that the C_{GD} component in symmetric and asymmetric structures contributes much less than the C_{GS} component to the total gate capacitance. For higher permittivity (k) in the SymD-k structure, C_{GS} increases rapidly with L_{hk}; which results in higher ON-current as shown in Figure 3.3. However, the delay performance of a logic circuit is directly dependent on the Miller component of C_{GD}. A higher value of C_{GD} in the high-k device severely degrades the delay and switching performance. Interestingly, in the SymD-k device, C_{GD} remains almost constant up to an optimal L_{hk} and then starts increasing because of stronger gate–drain coupling; whereas, the C_{GD} component remains almost the same throughout L_{hk} in the AsymD-kS device structure (due to the absence of drain-side high-k spacer). Therefore, the AsymD-kS structure would show better delay performances in comparison to the SymD-k structure that will be dealt with in the next section.

For digital applications, the device performance in terms of circuit delay depends on the relative rate of change of I_{ON} and C_{GG} (Sachid et al. 2008). Therefore, in most of the literature, the circuit delay performance is usually predicted by the relative rate of change of I_{ON} and C_{GG}. For achieving the substantial reduction in delay, I_{ON}/C_{GG} should be high enough.

Figure 4.6 shows the variations of normalized I_{ON}/C_{GG} as a function of L_{hk} for different k values. For the SymD-k FinFET, it is observed that an optimum I_{ON}/C_{GG} is obtained with L_{hk} ranging from 8 to 12 nm for different high-k spacer materials, which strongly impacts the delay of a circuit. However, in the case of the AsymD-kS FinFET, the I_{ON}/C_{GG} continuously decreases with an increase in L_{hk}, as shown in Figure 4.6b. This anticipates that the proposed AsymD-kS structure is not suitable for logic circuit applications in achieving better delay performances. Conversely, it is observed that the AsymD-kS structure outperforms the SymD-k and conventional one in terms of delay performance. Even the circuit delay reduction is more prominent with an increase in spacer permittivity. Note that the intrinsic advantages of the FinFET would not be visible if a simple $C_{GG}V_{DD}/I_{ON}$ delay metric is used for circuit performance evaluation. For a first-hand approximation, it helps to roughly anticipate the device dynamic performance in symmetrical architectures only. However, it may fail in predicting the dynamic performance for S/D asymmetric architectures.

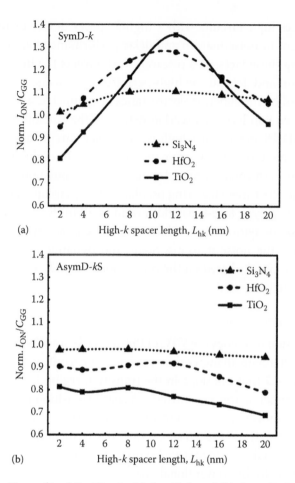

FIGURE 4.6 Normalized I_{ON}/C_{GG} in (a) SymD-*k* and (b) AsymD-*k*S structures for different spacer materials as a function of L_{hk}.

4.3 DUAL-*k* FinFET-BASED CMOS LOGIC PERFORMANCE: STABILITY AND SPEED

The primary goal in the CMOS logic circuit is to maximize stability and switching speed, simultaneously. However, the high-*k* structure with higher spacer permittivity could be beneficial in enhancing stability, but at the cost of an exorbitant increase in fringe capacitance; that in turn, degrades the delay performance. Therefore, most of the researchers have been forced to design logic circuits with low-permittivity spacers. To the best of our knowledge, no published material exists that designs and analyzes logic circuits with a high-k spacer device demonstrating improvement in noise margin and delay performances, simultaneously. Considering these facts, this research work targets the improvement in noise margin and delay using high-permittivity spacers.

Motivated by the superior electrostatics of symmetric and asymmetric dual-k spacer FinFETs, this section describes the suitability of SymD-k and AsymD-kS devices for high-performance circuit applications for improving the static and dynamic performances. To evaluate the FinFET logic circuit performance, mixed-mode circuit simulations of a FinFET inverter and the three-stage ring oscillator (RO3) circuits have been carried out. The p-type to n-type width ratio of a FinFET inverter is tuned to 2:1 to obtain symmetrical voltage transfer characteristics (VTC) and to maximize the noise margins. Static and dynamic characteristics of FinFET logic circuits based on the conventional, dual-k, and purely high-k structures are compared.

4.3.1 Static Performance: Noise Margins

This subsection discusses the VTC of the SymD-k– and AsymD-kS–based CMOS inverter with an increase in inner spacer permittivity. It is observed from Figure 4.7 that the slope in the transition region increases with an increase in inner-spacer permittivity. A sharp transition region improves the noise margins (NM_H, NM_L) of an inverter; that in turn, enhances stability. Accordingly, Figure 4.8 shows the improvement in maximum voltage gain achieved with SymD-k, AsymD-kS, and high-k–based inverters in comparison to the conventional one.

It is observed that for lower spacer permittivity (Si_3N_4, $k = 7.5$), both high-k– and SymD-k–based inverter shows similar improvement. However, with higher spacer permittivity materials, SymD-k not only outperforms the conventional and high-k but the AsymD-kS as well. In comparison to the conventional one, a SymD-k–based CMOS inverter shows an improvement by a factor of 4.1 and 7.6 in maximum voltage gain with an inner spacer permittivity of HfO_2 and TiO_2, respectively. These improvements are attributed to the very high drive current and better SCE control. The transition region slope is also dependent on the DIBL and subthreshold swing of a device. Therefore, the SymD-k structure performs better on all these grounds. Generally, it appears that the circuit stability has a negative correlation with DIBL (Song et al. 2010). Moreover, it is observed from the analysis carried out that the spacer

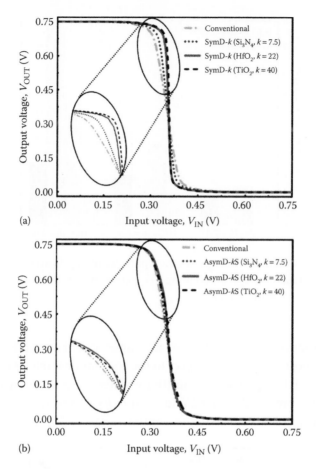

FIGURE 4.7 Voltage transfer characteristic comparison of (a) SymD-*k* and (b) AsymD-*k*S structures for different spacer permittivities.

permittivity (*k*) has a direct correlation with the inverter stability. In agreement to this, the NMs are considerably improved using higher inner permittivity in dual-*k* structures.

4.3.2 Dynamic Performance: Delay

This section describes the delay metric outcomes based on the choice of the high-*k* spacer material and its length in FinFET logic circuit applications. To evaluate the dynamic performance, a mixed-mode circuit simulation of the CMOS inverter and an RO3 circuit has been carried out.

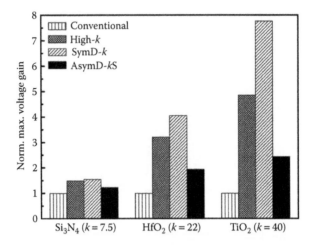

FIGURE 4.8 Maximum voltage gain comparison among SymD-k, AsymD-kS, and high-k structures for different spacer permittivities (k) normalized with respect to the conventional one.

4.3.2.1 CMOS Inverter

Figure 4.9 shows the inverter delay of symmetric and asymmetric dual-k structures normalized with respect to the conventional one as a function of L_{hk} (with HfO$_2$ as spacer permittivity material). It is clearly observed that both the proposed SymD-k and AsymD-kS architectures show better inverter delay performances with an optimized L_{hk} of 12 nm; even though,

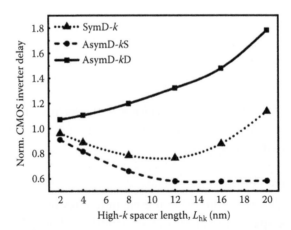

FIGURE 4.9 CMOS inverter delay for symmetric and asymmetric dual-k architectures as a function of L_{hk}, normalized with respect to the conventional counterpart.

both the structures also exhibit larger fringe capacitances. This superior delay performance is primarily due to the optimized high-*k* spacer length and its placement that modulates the field dynamics; and hence, electrostatics of the architecture.

It is discussed in the previous section that the SymD-*k* structure exhibits better static characteristics than the AsymD-*kS*. However, in terms of dynamic performance, the AsymD-*kS* outperforms the SymD-*k* architecture due to higher C_{GS} and smaller C_{GD} values. In a SymD-*k* device, the C_{GD} component sharply increases beyond an optimum L_{hk} of 12 nm; moreover, the C_{GS} reduces, which degrades the delay performance. On the other hand, the inverter delay worsens in the AsymD-*kD* device because of the lower C_{GS} and a high value of C_{GD}. Because of inferior performance, the subsequent sections will not deal with the AsymD-*kD* device.

Figure 4.10a and b plot the SymD-*k*– and AsymD-*kS*–based inverter delay, respectively, normalized with respect to the conventional one as a function of L_{hk} with different spacer permittivity materials. It is observed from Figure 4.10a that the inverter delay reduces with an increase in L_{hk} up to an optimal value of 12 nm; thereafter, it starts increasing sharply. For a purely high-*k* device (at L_{hk} = 20 nm), the inverter delay worsens by 15% and 27%; contrastingly, the SymD-*k*–based inverter (L_{hk} = 12 nm) speeds up by 24% and 32% for spacer materials of HfO_2 and TiO_2, respectively. As discussed earlier, the worst delay performance shown by the high-*k* device is because of the increased Miller capacitance (C_{GD}) beyond L_{hk} = 12 nm. Furthermore, it is also observed that the rate of decrement (or increment) in inverter delay in the vicinity of L_{hk} = 12 nm sharply increases with an increase in inner spacer permittivity.

Contrastingly, in the AsymD-*kS* structure (shown in Figure 4.10b), the inverter delay sharply reduces with an increase in L_{hk} at the source side up to an optimal value of 12 nm, and thereafter, it saturates. In comparison to the conventional one, the AsymD-*kS*–based inverter (at L_{hk} = 12 nm) speeds up by 42% and 54.4% for spacer materials HfO_2 and TiO_2, respectively. For similar delay improvement, AsymD-*kS* can also be considered with L_{hk} equal to L_{EXT}, i.e. 20 nm; which also simplifies the process complexity. However, it will slightly deteriorate the device as well as SRAM performance. Therefore, a better circuit performance of symmetric and asymmetric dual-*k* structures in terms of inverter delay and stability is obtained at L_{hk} = 12 nm with higher spacer permittivity.

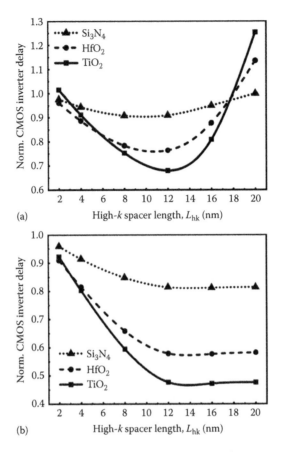

(a)

(b)

FIGURE 4.10 CMOS inverter delay for (a) SymD-k and (b) AsymD-kS architectures as a function of L_{hk} normalized with respect to the conventional one.

4.3.2.2 Three-Stage Ring Oscillator

For the accurate delay assessment, the suitability of the SymD-k and AsymD-kS FinFETs is also investigated by realizing an RO3 circuit and further compared with the conventional and purely high-k–based RO3 delay. The schematic of a tied-gate RO3 circuit is shown in Figure 4.11, in which each stage of an RO3 is a static FinFET inverter. Herein, V_1, V_2, and V_3 represent the output node voltages of the corresponding inverters.

The output waveforms of the conventional, SymD-k-, and AsymD-kS–based RO3 (Figure 4.12) show the stable frequency oscillations. The output frequency of a three-inverter-stage ring oscillator is obtained as $1/(6×$ inverter delay). Thus, the propagation delay of an inverter circuit is obtained by measuring the time period (T) of the oscillator.

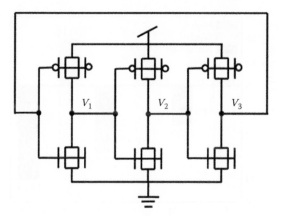

FIGURE 4.11   Schematic of the CMOS-based three-stage ring oscillator circuit.

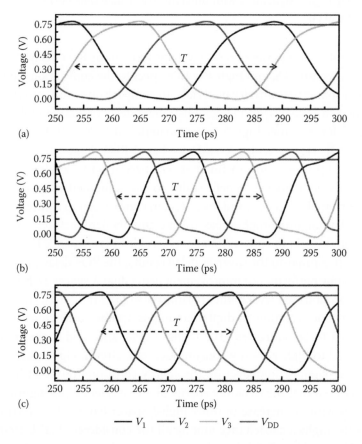

FIGURE 4.12   Output waveforms of the (a) conventional, (b) SymD-*k*, and (c) AsymD-*k*S FinFET-based three-stage ring oscillator.

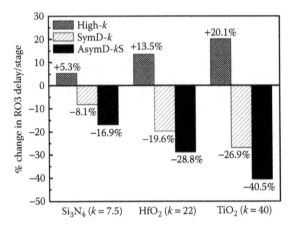

FIGURE 4.13 Percentage change in RO3 delay/stage using SymD-k, AsymD-kS, and purely high-k structures with different spacer materials normalized to the conventional one.

Figure 4.13 shows the percentage improvement in RO3 delay/stage for SymD-k, AsymD-kS, and high-k with respect to the conventional one for different spacer permittivity materials. In comparison to the conventional one, the SymD-k–based RO3 demonstrates a delay reduction of 8%, 20%, and 27% for an inner high-k spacer material of Si_3N_4, HfO_2, and TiO_2, respectively. However, the AsymD-kS–based RO3 demonstrates a delay reduction up to 41% for an inner spacer permittivity ranging from 7.5 to 40. This performance benefit is due to the combined effect of higher drive current, higher C_{GS}, and lower C_{GD} in comparison to a purely high-k device. Although the high-k–based logic circuits demonstrate better stability with respect to conventional one, its delay performance is the worst among all. Similar to the inverter delay results, the RO3 delay proves that the dynamic performance improvement using dual-k spacer architectures is more pronounced with higher permittivity of the inner spacers.

Energy consumption in digital circuits is another important concern that needs to be addressed. The higher capacitance and higher current would lead to higher energy consumption of a circuit. Figure 4.14 shows the spacer engineering effect on the energy consumption of an inverter. It is observed that the dynamic energy of an inverter circuit significantly increases with an increase in inner high-k spacer length. The AsymD-kS shows the highest capacitance among all the considered dual-k architectures. Hence, AsymD-kS–based inverter shows highest dynamic energy consumption as compared to others.

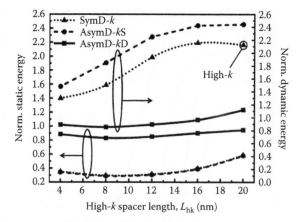

FIGURE 4.14 Static and dynamic energy consumption of an inverter with varying L_{hk}.

Similarly, AsymD-*k*D shows almost the same dynamic and static energy in comparison to the conventional case because of its reduced capacitance as well as current characteristics. It is also observed that the proposed symmetric dual-*k* (SymD-*k*) tri-gate–based inverter also consumes less dynamic energy in comparison to purely high-*k* devices. Moreover, from the static energy perspective, dual-*k* architectures outperform not only the high-*k* devices but also the conventional one due to the significantly lower leakage currents. It is a well-known fact that with the higher device density and lower supply/threshold voltage; the energy optimization focus has shifted from dynamic to leakage (static) energy while maintaining sufficient speed characteristics. The leakage energy is now dominating in the dense cache memories that occupy the major portion of a die. Therefore, we believe that for highly dense memories, the proposed SymD-*k* and AsymD-*k*S architectures would outperform both the conventional and purely high-*k* structures in terms of overall energy consumption.

4.4 EFFECT OF SUPPLY VOLTAGE ON DUAL-*k*–BASED INVERTERS

The power supply scalability is an important metric that needs to be explored from circuit perspectives. Figure 4.15 presents the effect of lowering the V_{DD} on VTC. The insets in Figure 4.15 show the voltage-gain comparisons of the dual-*k* FinFET-based CMOS inverter over the conventional one. Compared to the conventional, the SymD-*k*–based inverter shows prominent improvement in voltage gain with scaling of supply

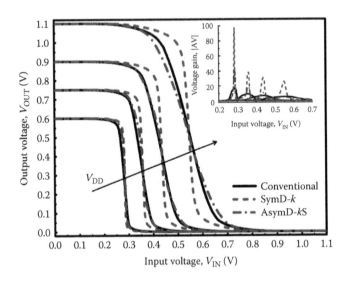

FIGURE 4.15   Effect of V_{DD} scaling on CMOS inverter VTC and voltage gain.

voltage. However, it is observed that the VTC of the AsymD-kS–based inverter degrades at higher V_{DD}.

It is because of the enhanced GFIBL effect at higher V_{DD} that significantly reduces the G/S underlap barrier. Therefore, the CBE barrier (in the ON-state) directly under the gate becomes prominent (instead of the underlap barrier) limiting the drive current in AsymD-kS; and hence, the static performance.

4.5 SUMMARY

This chapter comprehensively analyzed the role of fringe capacitances associated with proposed dual-k architectures that advocates for high-k spacer materials for improving noise margin and delay performances, simultaneously. The dual-k structures exhibit larger fringe capacitances, but with an optimized inner spacer length of 12 nm; both the SymD-k and AsymD-kS architectures show better inverter delay performances. This superior delay performance is primarily due to the diligent usage of high-k spacer length and its placement that modulates the field dynamics; and hence, electrostatics of the architecture. The AsymD-kS outperforms the other architectures in terms of dynamic performances due to higher C_{GS} and smaller C_{GD} values. Moreover, an important and novel observation is made that the delay performance improvement is more pronounced with higher permittivity of the spacers in dual-k spacer technology that otherwise worsens in the case of purely high-k architecture. Furthermore, this chapter also investigated the effect of power supply scalability on dual-k–based circuits.

REFERENCES

Agrawal, S. and J. G. Fossum, A physical model for fringe capacitance in double-gate MOSFETs with non-abrupt source/drain junctions and gate underlap, *IEEE Trans. Electron Dev.*, 57(5), 1069–1075, 2010.

Colinge, J. P., *FinFETs and other Multi-Gate Transistors*, New York: Springer-Verlag, 2008.

Dey, A., A. Chakravorty, N. Dasgupta, and A. Dasgupta, Analytical model of subthreshold current and slope for asymmetric 4-T and 3-T double-gate MOSFETs, *IEEE Trans. Electron Dev.*, 55(12), 3442–3449, 2008.

Kim, S. H., J. G. Fossum, and J. Yang, Modeling and significance of fringe capacitance in nonclassical CMOS devices with gate-source/drain underlap, *IEEE Trans. Electron Dev.*, 53(9), 2143–2150, 2006.

Majumdar, K., R. S. Konjady, R. T. Suryaprakash, and N. Bhat, Underlap optimization in HFinFET in presence of interface traps, *IEEE Trans. Nanotechnol.*, 10(6), 1249–1253, 2011.

Manoj, C. R. and V. R. Rao, Impact of high-k gate dielectrics on the device and circuit performance of nanoscale FinFETs, *IEEE Electron Dev. Lett.*, 28(4), 295–297, 2007.

Mohapatra, N., M. P. Desai, S. G. Narendra, and V. R. Rao, The effect of high-k gate dielectrics on deep submicrometer CMOS device and circuit performance, *IEEE Trans. Electron Dev.*, 49(5), 826–831, 2002.

Sachid, A. B., C. R. Manoj, D. K. Sharma, and V. R. Rao, Gate fringe-induced barrier lowering in underlap FinFET structures and its optimization, *IEEE Electron Dev. Lett.*, 29(1), 128–130, 2008.

Schulz, T., C. Pacha, and L. Risch, Impact of technology parameters on device performance of UTB-SOI CMOS, *Solid State Electron.*, 48(4), 521–527, 2004.

Shahrjerdi, D., J. Nah, T. Akyol, M. Ramon, E. Tutuc, and S. K. Banerjee, Accurate inversion charge and mobility measurements in enhancement-mode GaAs field-effect transistors with high-k gate dielectrics, *Device Res. Conf.*, 29, 73–74, 2009.

Sharma, R. K., M. Gupta, and R. S. Gupta, TCAD assessment of device design technologies for enhanced performance of nanoscale DG MOSFET, *IEEE Trans. Electron Dev.*, 58(9), 2936–2943, 2011.

Shenoy, R. S. and K. C. Saraswat, Optimization of extrinsic source/drain resistance in ultrathin body double-gate FETs, *IEEE Trans. Nanotechnol.*, 2(4), 265–270, 2003.

Song, X., M. Suzuki, T. Saraya, A. Nishida, T. Tsunomura, S. Kamohara, K. Takeuchi, S. Inaba, T. Mogami, and T. Hiramoto, Impact of DIBL variability on SRAM static noise margin analyzed by DMA SRAM TEG, in *Technical Digest IEDM*, IEEE, San Francisco, CA, pp. 62–65, 2010.

Trivedi, V., J. G. Fossum, and M. M. Chowdhury, Nanoscale FinFETs with gate-source/drain underlap, *IEEE Trans. Electron Dev.*, 52(1), 56–62, 2005.

Vellianitis, G., M. van Dal, L. Witters, G. Curatola, G. Doornbos, N. Collaert, C. Jonville, et al., Gatestacks for scalable high-performance FinFETs, in *IEDM Technical Digest*, IEEE, Washington, DC, pp. 681–684, December 2007.

CHAPTER 5

Design Metric Improvement of a Dual-*k*–Based SRAM Cell

5.1 INTRODUCTION

The continuous increase in the dataset size and wide gap between the speed of processors and main memories has led to an ever-increasing demand for large cache memories. In modern processors, the caches account for a significant fraction of the chip area as well as the power consumption. Recently, the ITRS predicted that by the end of the year 2016, the memory circuits would occupy 94% area of a chip (ITRS 2013). Traditionally, SRAM has been the workhorses for realizing cache memories due to its robustness, relatively lower read/write access times, and process compatibility. Most of the time the large SRAM cell array remains idle; therefore, static power consumption is a big issue for memory circuits. However, exorbitantly high leakage currents and short-channel effects such as V_{th} roll-off, drain-induced barrier lowering (DIBL), and subthreshold slope pose several significant challenges to SRAM design. To tackle the leakage problem and to improve SRAM stability, earlier research efforts have explored a number of different device and circuit-level techniques such as multi-V_{th} and multi-t_{ox} transistors (Rostami and Mohanram 2011), body/back-gate biasing (Giraud et al. 2009), write assist (Yang et al. 2012), and 8T/10T bit-cells (Kanj et al. 2008). However, as the technology scales down to sub-20 nm nodes, extremely small channel lengths and close proximity between highly doped source and drain regions introduce newer leakage components such as the direct source-to-drain tunneling leakage in addition to subthreshold thermionic leakage and gate leakages (Vega and Liu 2010). This leads to very high leakage currents as well as degradation of the transistor I_{ON}/I_{OFF} ratio, which impacts the leakage and access time of the cell. The other major design concern is the read/write conflict; wherein, a transistor sizing to enhance the read stability degrades the write ability and vice versa. To enhance read stability, a cell ratio (CR) must be increased either by increasing pull-down transistor width or by increasing access transistor length. Both options degrade write ability, area as well as power dissipation (Giraud and Amara 2008). Similarly, to enhance write ability, a pull-up ratio (PR) must be decreased. So, the main conflict arises due to the current driving capability of the access transistor.

As the FinFET offers some unique features, solutions can be achieved by exploiting these features. Several radical departures from conventional design have been discussed earlier in Section 2.6.3 that claimed the mitigation of read–write stability conflict, improved noise margins, and access times. A substantial volume of research focused on independent-gate configurations for threshold (V_{th}) adjustment (Giraud et al. 2009) that also enhanced the circuit complexity. Recently, many innovative source/drain asymmetric architectures have been proposed to mitigate read/write conflict and to improve the SRAM cell metrics. Source/drain asymmetry helps in adjusting the PR and CR to augment the stability; but in turn, adversely affects the cell area, leakage power, and access times. Goel et al. (2011) proposed an asymmetric structure that enhances read/write stability and reduces leakage at the expense of higher access time and cell area. Moradi et al. (2011) proposed an asymmetrically doped (AD) FinFET structure that reported an improvement in static noise margins (SNMs) and cell leakage but with a higher access time penalty. Recently, Sachid and Hu (2012) proposed a stable SRAM cell structure by fabricating multiple fin heights (H_{fin}), that is possibly a better solution to the problem of width quantization. However, it is an application-specific approach with the increased process complexity. Moreover, all these structures would require setting up of new design rule constraints, thus limiting their applications.

Motivated by the superior device electrostatics and circuit performance, this chapter investigates the proposed symmetric (SymD-k)

and asymmetric dual-k (AsymD-k) architectures in the 6T-SRAM cell. It is observed that the proposed SymD-k–based SRAM cell helps in improving the hold, read and write noise margins, read/write speed (access time), and standby leakage power without affecting PR and CR. Moreover, the AsymD-k–based SRAMs also enhance stabilities, write delay, and leakage power without cell area and read delay penalties. The rest of the chapter is organized as follows. Section 5.2 presents the brief introduction of a basic SRAM architecture and bit cell, its read/write operations, and the performance evaluation metrics such as SNM, access time, and standby leakage power. Thereafter, Section 5.3 explores the proposed symmetric and asymmetric dual-k configurations, and their merits and demerits over the conventional and purely high-k architectures. In Section 5.4, the effect of supply voltage on dual-k–based SRAM cells is investigated. Section 5.5 compares both the proposed dual-k–based SRAM cells on a common platform by evaluating power margins. At last, Section 5.6 draws a brief summary and the major outcomes of this chapter.

5.2 FUNDAMENTALS AND EVALUATION METRICS FOR AN SRAM CELL

An SRAM cache consists of an array of bistable memory bit cells along with the address (row and column) decoders, sense amplifiers, write drivers, and bit-line precharge circuits commonly known as the peripheral circuitry. Peripheral circuitry helps in reading from and writing into the array. A conventional SRAM array is composed of millions of identical cells. For example, a 32 Mb cache memory is composed of 33,554,432 cells. Each cell circuit is capable of storing a single bit of information. Therefore, a small improvement in reliability, performance, and saving in static power will strongly impact the entire processor. As most of the chip area in a cache is covered by the bit-cell component, several researchers targeted to improve SRAM cell performance in the past four decades (Ananthan and Roy 2006; Bansal et al. 2007; Cheng et al. 2007; Giraud et al. 2009; Bhattacharya and Jha 2014).

A standard tied-gate 6T SRAM cell consists of two cross-coupled inverters and two access transistors connected to each data storage node. The schematic circuit of the same is shown in Figure 5.1 with different leakage current contributions. The inverter pair forms a latch and holds the binary information. The data in the SRAM cell are stored as long as the

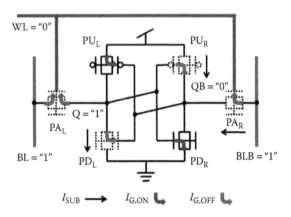

FIGURE 5.1 A standard tied-gate 6T SRAM cell schematic with different leakage current contributions.

power is ON. The cross-coupled inverters are connected to two bit lines, commonly known as BL (bit line) and BLB (bit line bar/complement), through access transistors PA_L and PA_R, respectively. The access transistors are controlled by the word-line (WL) voltage. The SRAM cell has three modes of operation: read, write, and standby. In other words, it can be in three different states such as reading, writing, or standby. In the standby or hold mode, WL is kept low ($V_{WL} = 0$ V), thus turning OFF the access transistors and isolating the bit lines from the cross-coupled inverter pair.

In a read operation, prior to the WL being selected (V_{WL} raised from 0 to V_{DD}), both the bit lines are precharged to V_{DD} via the low-impedance path. Then, the selected WL is enabled ($V_{WL} = V_{DD}$), activating the access transistors of the desired row. By turning ON the access transistors of a row, a small difference voltage is generated between each bit-line pair, connected to it. This small voltage difference is detected and amplified by the sense amplifiers connected to the bit lines. At the end of the read operation, the WL is turned OFF; thus, isolating the cell from the bit lines and allowing data nodes to return to their standby values before the read cycle. Proper design needs to be exercised to ensure that the values on storage nodes are not flipped during the read operation. However, in a write operation, appropriate write voltages are applied to the bit lines to force the cell into the intended logical state. A write operation starts by applying a voltage at bit line that corresponds to the data to be stored in the cell. Then, the WL is enabled and the memory cell flips to the state corresponding to the voltage difference. The write operation is completed by turning OFF the WL by a column decoder. Throughout the chapter, the cell performance is mainly evaluated on the basis of SNMs (hold, read, and write), read/write access time, and the standby leakage power. The traditional measure to find the stability of an SRAM cell during the different modes of operation is the SNM.

The SNM is defined as the maximum amount of DC noise voltage that can be tolerated by the cross-coupled inverter pair such that the cell retains its data (Seevinck et al. 1987). Both the hold and read SNMs are extracted from the voltage transfer characteristics (VTC) of hold and read operations, respectively. During the hold/retention mode, nodes Q and QB store logics "1" and "0," respectively, and WL is OFF. Hold SNM defines the stability in retaining the stored data. However, during a read operation, the WL access transistors are ON after bit lines are precharged. The read VTC can be measured by sweeping the voltage at the data storage node Q (or QB) with both the bit lines (BL, BLB) and WL biased at V_{DD}, while monitoring the node voltage at QB (or Q). For higher read stability, access transistor strength must be low. The access time in the read mode is another important metric, which depends on read cell current through the access and pull-down transistors. Similarly, write margin and write access time are calculated for write operation. The write access time is measured between the time when WL reaches to 50% of V_{DD} and node QB reaches the switching threshold voltage of the other inverter.

5.3 DESIGN CHALLENGES OF FinFET-BASED SRAMs

FinFET is an emerging technology. On the other hand, the memory circuits would occupy 94% area of a total chip (ITRS 2013). Therefore, it is very important to have an overall literature review to understand the progress of FinFET-based SRAM cells considering the device, circuit, and technological issues.

Designing a power-efficient, area-efficient, and robust SRAM cell is a major concern. Furthermore, there are also several important trade-offs that need to be considered to find an optimized design for SRAM. Most of the time the large SRAM cell array remains idle; therefore, the static power consumption is a big issue for memory circuits. To reduce the leakage, there are two techniques that can be followed. The use of longer channel length is one of the two techniques, but it further affects the area, capacitance, access time, and active power. Another technique is to use higher threshold voltage; that in turn, affects the access time. The larger size transistors ensure larger design margins; therefore, it offers enhanced device performance but at the cost of area. To reduce the area, a lower supply voltage can be used that also reduces the leakage power consumption.

In addition, it degrades the SNM. The increase in gate leakages and the decrease in the I_{ON}/I_{OFF} current ratio also affect the SNMs. Access time is an important parameter for SRAM cell design and it is dependent on the successful read/write operation (Guo et al. 2005; Tawfik and Kursun 2008). Due to less channel doping and larger effective channel width, it is possible to achieve lower operating voltage; and therefore, less dynamic power in FinFET SRAM circuits. The better control over the channel due to double gate allows us to achieve better SCEs with low static power consumption.

As we already discussed, the fin width can only be increased in quanta of fin height. Therefore, the width quantization became a major issue for FinFET-based SRAM circuits. The SRAM circuits require proper transistor sizing for the robust performance and can be achieved through width optimization. However, width optimization is difficult in FinFET compared to the planar technology. Moreover, the conflicting design requirements in the 6T SRAM cell for achieving high read and write stabilities make the situation even more complicated. Considering these facts, Kim and Fossum (2007) reported the design optimization and performance projections of double-gate FinFETs with gate-source/drain underlap for SRAM applications.

5.4 REVIEW OF TECHNOLOGY-CIRCUIT COOPTIMIZATION OF FinFET-BASED SRAMs

Considering the above challenges, Kim and Fossum (2007) reported the design optimization and performance projections of double-gate FinFETs with gate-source/drain underlap for SRAM applications.

5.4.1 Source/Drain Asymmetric FinFET-Based SRAM Cells

In the reported literature, several radical departures such as S/D asymmetry, back-gate biasing, and usage of some performance boosters in FinFETs have been proposed earlier that claimed the mitigation of read–write stability conflict. A substantial volume of research focused on the independent-gate configurations for threshold (V_{th}) adjustment (Masahara et al. 2007; Endo et al. 2008; Gupta et al. 2011), while several others proposed asymmetric S/D device architectures (Goel et al. 2011; Moradi et al. 2011) that enhanced the process as well as circuit complexity. The S/D asymmetry helps in altering the PR and CR exploiting the bidirectional current flow to augment the SNMs; but in turn, adversely affects the cell-area, leakage power, and access times. In this subsection, we analyze the concept of introducing asymmetry in the FinFET device and the way it affects the SRAM cell performance and robustness.

Goel et al. (2011) presented an asymmetric drain spacer extension that introduced a gate underlap only on the drain side using an extended spacer. Compared to the conventional FinFET SRAM, the asymmetric FinFET exploits the magnitude of currents for positive and negative drain-to-source voltages. The authors have claimed to achieve a 57% decrease in leakage, 11% improvement in the read SNM, and 6% improvement in the write margin. However, it suffered from degraded access time (7%) and cell area (7%). Similarly, Moradi et al. (2011) proposed an AD FinFET in which asymmetry in the device is achieved by unequal source/drain doping of FinFETs. On the basis of this, the authors designed a FinFET SRAM cell that would simultaneously improve read and write margins and also mitigate the read/write conflict. Using an AD FinFET, they achieved superior short-channel characteristics, lower cell leakage, improved read SNM, write margin, write time, and hold SNM. This AD-based SRAM cell is also able to resolve the read–write conflict as the strength of the access transistors varies with the storage node voltage. The improvements reported in read and write SNMs are 7.3% and 23%, respectively. However, these improvements also come at the cost of an excessively increased access time of 42% due to weak access transistor during the read operation. Recently, Sachid and Hu (2012) proposed a stable SRAM cell structure by fabricating multiple fin heights that might be a better solution to the problem of width quantization, however, with an increased process complexity.

Ebrahimi et al. (2012) studied the different characteristics of 6T and 4T SRAM cells based on asymmetric FinFET structures. Recently, Salahuddin et al. (2013a) proposed two novel 6T FinFET SRAM cells based on an asymmetrical gate underlap technique under process parameter fluctuations. In the first design, they constructed the cell with asymmetrically gate overlap/underlap engineered FinFET-OU. In this design, the right side of the device is underlapped and the left side is overlapped. In this SRAM-OU cell, the underlapped sides of asymmetrical FinFET-OU are connected to the data storage nodes and the overlapped sides are connected to the bit lines. In their second design, Salahuddin et al. (2013b) proposed a memory cell based on asymmetrically gate underlap engineered FinFET-AU. Here, the longer underlapped sides of asymmetrical FinFET-AUs are connected to the data storage and the shorter underlapped sides are connected to the bit lines. As the direction of the current flow is reversed, the strengths of the asymmetrical bit-line access transistors are weakened during read operations and enhanced during write operations. Therefore, the conflicting design requirements of achieving read/write stability can be mitigated. Both the proposed asymmetrical designs provide a much better read SNM of around 70% compared to symmetrical gate underlap-based design. However, the leakage power was reduced with the second design compared to the first one. Hu et al. (2010) reported that the asymmetric source-underlap access transistors can improve the read SNM but can degrade the write margin. They observed that the FinFET SRAM cell based on asymmetric source/drain underlap access and pull-up transistors (at $V_{DD} = 1$ V) can improve the Read Static Noise Margin (RSNM) by 20.5% with the comparable Write Static Noise Margin (WSNM), 10% reduction in cell read access time, and 36% improvement in write time as compared to a symmetrical FinFET SRAM cell. However, due to the worse electrostatic integrity caused by the underlap; it is not possible to further improve the RSNM using source/drain-underlap access transistors as the V_{DD} is reduced below 0.6 V.

5.4.2 Independent-Gate–Based FinFET SRAMs

As the circuit designs with FinFETs offer some unique features to explore, the independent-gate FinFET is likely to be the most important one. For the independent-gate–based FinFET designs, one gate can be used for driving/switching and the other gate can be used for threshold voltage control. It allows the dynamic threshold voltage control due to the electrical coupling of both the front and back gates. The modulation of front-gate threshold voltage can be achieved by applying back-gate voltage biasing. Therefore, it offers more stable SRAM circuits. Also, it is possible to reduce the trade-offs between the read and write margins by applying this technique. Thus, the improved device performance, reduced leakage current, and stability improvement can be easily achieved by using an independent-gate FinFET. Cakici and Roy (2007) showed that, a further control on leakage can be achieved using the sleep transistor-based source biasing technique with independent-gate–based FinFET SRAM. Theoretically, it is preferred that back-gate devices should be asymmetrically built. Joshi et al. (2007) described the area-efficient row-based back-gate biasing scheme for an asymmetrical double-gate FinFET. On the basis of the comparisons, they proved that the FinFET-based back-gate biasing to control the V_{th} of the device was better than the CMOS body/well biasing. Later on, Kanj et al. (2008) presented a column-decoupled SRAM design and revealed the statistical evaluation and yield. Two novel independent-gate FinFET-based SRAM cells were presented by Tawfik and

Kursan (2007). In their first design, the pull-down transistors are tied-gate transistors; whereas, the pull-up transistors and the access transistor are independent-gate FinFETs. By this approach, a 50% enhancement in SNM has been reported compared to the tied-gate FinFET SRAM cell. In the second design, the technique of V_{th} modulation has been explored by using an IG access transistor by dynamically tuning the read/write strength. For read and write operations, they proposed two separate data access mechanisms. The back gate of access transistors is controlled by a separate write signal (W), and the front gate is controlled by a read/write (R/W) signal. The threshold voltage of the access transistor is dynamically adjusted using this technique. The read SNM has been enhanced by 92% with this approach. The leakage power reduction is reported as 36% using this IG approach compared to the tied-gate approach. Liu et al. (2008) reported a 17.5% reduction in area compared to the tied-gate FinFET using minimum-sized independent-gate transistors without affecting the data stability. Endo et al. (2008) proposed a row-by-row V_{th} control for an independent-gate-based SRAM array. In this design, back-gate control lines parallel to the word lines are used to control the V_{th}. In standby mode, the threshold voltage of the transistors was increased to reduce the leakage current. During a read or write access, the threshold voltage was decreased to ensure high drive current. The voltages for the control lines are supplied by level shifters by converting a row decoder output signal. Endo et al. (2008) fabricated a FinFET-based SRAM array and reported a drastic reduction in leakage power consumption along with dynamic power consumption by efficiently controlling the back gate.

5.4.3 Fin-Thickness, Fin-Height, and Fin-Ratio Optimization

FinFET offers several improvements in SRAM performance, but one needs to address many challenges also that arise due to inherently different device architectures. This subsection discusses the effect of fin thickness and fin height on DIBL, threshold voltage, write ability, and read stability of the SRAM cell. For superior circuit/SRAM performance, it is very important to choose the correct/optimized fin configuration. Here, fin configuration means fin height, thickness, fin pitch, the number of fins, and so on. The SRAMs are interconnect-dominated circuits; therefore, the increase in drive current would be beneficial for memory circuits. It is possible to increase the effective channel width and drive current by increasing the fin height of the FinFET. In a joint optimization study of V_{DD}, fin height, and V_{th} on SRAM performance;

Ananthan and Roy (2006) achieved 87% lower subthreshold leakage, 50% lower gate leakage, 25% lower dynamic energy, 13% higher SNM using 69% taller fins, 8% lower V_{DD}, and 35% higher V_{th}. Furthermore, they reported that the increase in fin thickness (T_{fin}) lowers the read stability of the FinFET SRAM cell. However, larger T_{fin} increases the yield of write ability or WL trip voltage. On the other hand, the read stability decreases as the fin height (H_{fin}) increases, but if the T_{fin} is small, then the H_{fin} does not have a significant impact on read stability (Lee et al. 2012). In one of the studies, Dobrovolny et al. (2012) analyzed the impact of interdie fin height variations on the SRAM cell; wherein, they reported that the interdie fin height variations dominate the overall intradie variation and both interdie variation and intradie variation affect the SRAM cell SNM and write-trip point. Chen et al. (2013) reported that a multiple fin height SRAM cell demonstrates a 25% better SNM than the single fin height cell. They used a tall fin in the pull-down transistor and a short fin in the pass gate transistor. Cheng and Li (2010) reported that a multifin and larger aspect ratio SRAM showed better electrical characteristics than a single fin and smaller aspect ratio FinFET. Also, a multifin FinFET reduces the random dopant fluctuations because of uniform surface potential. On the other hand, Kang et al. (2010) demonstrated that a two-fin pass gate consumed larger dynamic energy due to the effectively larger gate and drain capacitances. Therefore, a one-fin pass gate would be a suitable choice for SRAM designing with proper surface orientation. Another study on the process variation impact on the FinFET SRAM reported that the stability of the cell was most susceptible to the fin thickness variation of the access transistor (Shimeng et al. 2008). They demonstrated that using multiple fins in the pull-down transistor might improve the read SNM in the worst case. Sohn et al. (2013) stated that the parasitic capacitance can be lowered by using taller and denser fins. Also, the static power consumption was much larger with taller fins due to higher off-current. Rasouli et al. (2009) reported that the fin thickness below 10 nm (for 22 nm technology) can induce the quantum mechanical effect such as structural confinement. Matsukawa et al. (2009) demonstrated that the parasitic resistance is greatly influenced by T_{fin} fluctuations and can be reduced by optimizing the extension doping. An optimization for the stability of SRAM has been discussed through the silicon fin thickness and fin ratio by

Lekshmanan et al. (2007); wherein, they demonstrate that the silicon thickness constraint can be relaxed without affecting the stability of the cell and can reduce the process variability. Also, it has been reported that the reduction of the fin-to-fin variability can be achieved by increasing the number of fins. Besides this, a penalty in terms of 25% increase in area is paid on increasing the number of fins by one or two in access or pull-down transistors.

5.4.4 Fabrication Level Optimization

In the preceding subsections, it is clearly seen that the performance metrics of FinFET-based SRAMs can be improved using various device and circuit level approaches. In this subsection, we focus on fabrication level optimization techniques. Kawasaki et al. (2009) discussed the fabrication-related challenges and solutions in SRAM cells for 22 nm node and beyond. They reported that the sidewall image transfer (SIT) is an important technique to achieve narrower fin formation. Furthermore, this technique can be advantageous for SRAM circuits to achieve lower line edge roughness. Kawasaki et al. (2009) preferred to use plasma doping or solid-phase doping to reduce the threshold voltage mismatch in SRAMs. As we have discussed earlier, the reduction in parasitic resistance is a key factor in improving FinFET device performance. Therefore, Kawasaki et al. (2009) proposed a source/drain merged FinFET with diamond-shaped epi to improve the parasitic resistance issue. A high-performance 25 nm gate length SRAM cell has been fabricated in Chang et al. (2009) by optimizing fin extension and embedded SiGe source/drain. The optimized implant and anneal conditions can be used to reduce the extension resistance; that in turn, improves the fin quality and short-channel effects as well. Liu et al. (2006) reported that the TiN wet etching technique is good to obtain sub-30 nm gate length, taller fins of around 100 nm and symmetrical threshold voltages. Veloso et al. (2009) demonstrated a FinFET-based SRAM cell by using an advanced single-patterning process with full-field EUV and immersion lithography. To achieve good electrical characteristics at V_{DD} down to 0.4 V, the W metallization for contact holes and epitaxial raised source/drain (SEG) with double spacer and ultra-thin silicide were also demonstrated. In their later work, Horiguchi et al. (2010) reported that the single gate patterning approach produced a weak yield in the FinFET SRAM and had a problem in gate pitch control. Therefore, they demonstrated the double-gate patterning approach to improve the yield of the FinFET SRAM. Kawasaki et al. (2008) demonstrated a single-sided ion implantation scheme to reduce threshold voltage variation in the FinFET-based SRAM. Also, they reported that an undoped channel FinFET-based SRAM shows lesser threshold voltage variation compared to the halo-doped FinFET SRAM. Basker et al. (2010) demonstrated a FinFET SRAM cell operating at 0.4 V with good performance metrics. Although they used the conventional optical lithography, they managed to aggressively scale fin pitch (40 nm) and gate pitch (80 nm) by using a double-exposure, double-etch SIT process. Besides this, they also used the epitaxial films for conformal doping and reported that it reduces the external resistance by 30%.

5.5 DUAL-*k* SPACER ENGINEERED SRAM CELLS

This section describes the design metric outcome of SymD-*k* and AsymD-*k* tri-gate FinFET-based 6T SRAM cells. Figure 5.1 shows the schematic of a tied-gate 6T SRAM cell with all leakage current components. The thick line is used to represent the large line capacitances associated with the WL and bit lines (BL and BLB). The PMOS pull-up transistors (PU_L and PU_R) and NMOS access transistors (PA_L and PA_R) are of minimum size (single fin) to set a low PR value. All the analysis and comparisons are drawn (considering HfO_2 an inner high-*k* spacer) based on the simulations performed for a CR of two by modulating fin-pitch; and hence, current driving capability of pull-down transistors (PD_L and PD_R).

5.5.1 Symmetric Dual-*k* (SymD-*k*) FinFET-Based SRAM Cell

This subsection presents the SRAM metrics enhancement using the SymD-*k* FinFET structure in comparison to the conventional and high-*k* FinFET-based SRAMs. The primary goal in the SRAM cell includes maximizing stabilities and minimizing access times besides achieving minimum leakages. In general, the read and write stability depends on the resistive divider action of PA-PD and PU-PA transistors, respectively. The storage node voltage is charged or discharged through the access transistor. To prevent read failure in a cell, the storage node voltages must be less than the inverter trip voltage. Therefore, the PDs must be stronger than PAs. For reliable write operations, the PA transistor strength must be larger than the PU transistor strength. This contradictory sizing requirement of access transistor raises read/write conflict. Therefore, it is necessary to use improved architecture that exhibits superior device performance while mitigating read/write conflict with improved access times. To the best of our knowledge, none of the published material has discussed the collective improvement in the stability, access time as well as standby leakage power without affecting the design ratios (CR and PR) and cell area.

The static noise margin (SNM) comparisons of the proposed SymD-*k*– based SRAM cell with respect to the conventional and high-*k* based SRAM in all three possible modes of operations (hold, read, and write) are shown in Figure 5.2. The hold, read, and write margins are improved by 8.7%, 9.4%, and 10.4%, respectively, compared to the conventional SRAM cell. These improvements in SNMs are attributed to the enhanced electrostatic control that increases noise margins (NM_H and NM_L) and voltage gain of an inverter (discussed in Section 4.3.1). Also, the read/ write stability is not directly dependent on the absolute value of I_{ON} (Kim and Fossum 2007). Apparently, the SNM has a negative correlation with DIBL (Song et al. 2010). In agreement with this, it is observed that the SNMs are considerably improved using the proposed SymD-*k*

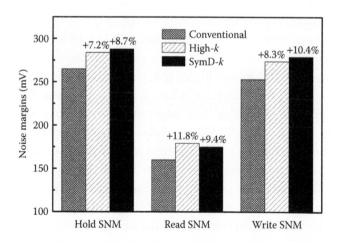

FIGURE 5.2 Comparison of hold SNM, read SNM, and write margin among the conventional, purely high-*k*–, and SymD-*k*–based 6T SRAM cells.

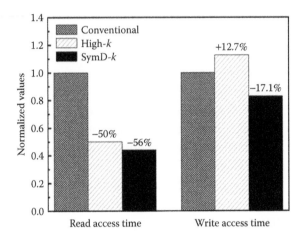

FIGURE 5.3 Comparison of read and write access times in purely high-k– and SymD-k–based 6T SRAM cells normalized with respect to the conventional one.

architecture without affecting the design ratios, that is, CR and PR and cell area penalty.

Furthermore, there are 2.31× and 1.22× reduction in read and write access times, respectively, compared to the conventional one, as depicted in Figure 5.3. The read access time depends on the read cell current through the PA transistors; therefore, the SymD-k cell outperforms the other two cells. It is also observed that the purely high-k SRAM cell proves to be better than the conventional one in all respects except for the 12.7% increase in write access time. The optimal high-k spacer length in the SymD-k structure increases I_{ON}/C_{GG}, which helps in reducing write access time; however, the purely high-k–based SRAMs show worst write delay among all. Although the purely high-k–based SRAM cell helps in stability enhancement and reduces read access time, it lacks in terms of write access time and leakage power in comparison to the proposed SymD-k SRAM cell.

Figure 5.4a–c show the percentage change in hold-SNM, read, and write access time, respectively, as a function of the inner high-k spacer permittivity and length. It is observed that the reduced read/write access time and enhanced SNMs can be achieved by using an inner spacer length ranging from 8 to 12 nm for $L_{un} = 8$ nm; however, beyond this range, the read access and hold SNM slightly degrade. On the other hand, increasing the value of k for L_{hk} outside the range of 10–15 nm degrades the write access time more

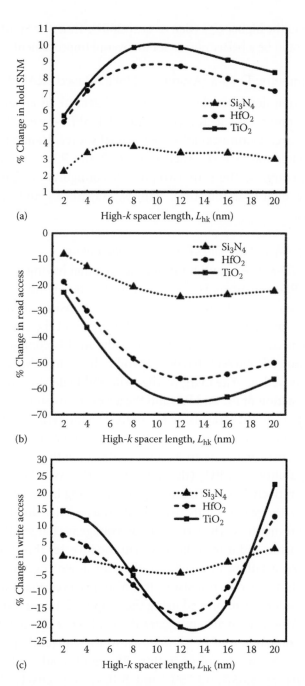

FIGURE 5.4 Percentage change in (a) hold SNM, (b) read access time, and (c) write access time with respect to the conventional FinFET-based SRAM for different high-*k* values as a function of L_{hk}.

rapidly due to the higher C_{GD} (Bansal et al. 2007). Thus, the SymD-k device structure would be a better option for an overall improvement.

5.5.2 Asymmetric Dual-k (AsymD-k) FinFET-Based SRAM Cell

This subsection describes the AsymD-k architecture to enhance SRAM metrics in comparison to the conventional FinFET-based SRAM. This asymmetric architecture exploits the unequal source/drain currents that help in modulating SRAM design ratios (CR and PR) to improve read/write noise margins. There are two possible configurations based on the asymmetric dual-k architecture, namely, AsymD-kS and AsymD-kD SRAM cells.

5.5.2.1 The AsymD-kD FinFET SRAM Configuration

In the AsymD-kD cell configuration, the terminal having dual-k spacer of PU and PD transistors is connected to the storage nodes for implementing cross-coupled inverters. The access transistor (PA) terminal with dual-k spacer is connected to the bit line, and the other terminal with low-k is connected to the storage node. In this configuration, the use of asymmetric current driving capability helps to enhance the hold, read SNM as well as write margin. The write ability depends on the resistive divider action of PU and PA, and the access transistor strength must be larger than the PU transistor strength. During write operation, the terminal having a dual-k spacer of PA_R is at higher voltage; therefore, the access transistor acts as an AsymD-kS device with very high current driving capability than PU_R (that operates as the AsymD-kD structure with comparatively lower drive strength), resulting in higher write ability. Similarly, the read stability depends on the resistive divider action of PA and PD transistors. The storage node voltage is charged through the access transistor, and it must be less than the inverter threshold voltage to prevent read failure in a cell. So, the pull-down transistor must be stronger than the access transistor. It is also observed that the read stability is considerably improved using the proposed AsymD-kD cell configuration. Moreover, it also mitigates the read/write conflict. Figure 5.5 shows the comparison of noise margins to the conventional SRAM in all three possible modes of operation. The hold, read, and write margins increase by 5.66%, 13.75%, and 5.16%, respectively. Figure 5.6 shows a 14.38% reduction in write access time due to the stronger access

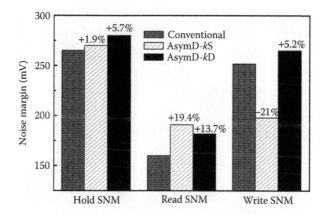

FIGURE 5.5 Comparison of hold, read, and write margins among the conventional, AsymD-*k*S–, and AsymD-*k*D–based 6T SRAM cells.

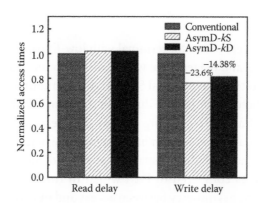

FIGURE 5.6 Comparison of read and write access times in AsymD-*k*S– and AsymD-*k*D–based 6T SRAM cells normalized with respect to the conventional one.

transistor during write mode and the same read access time in comparison to the conventional FinFET-based SRAM.

5.5.2.2 The AsymD-kS FinFET SRAM Configuration

In this SRAM cell configuration, the AsymD-*k*S FinFET structure is used. The terminals having dual-*k* spacer of pull-up and pull-down transistors are connected to V_{DD} and *GND*, respectively, to implement cross-coupled

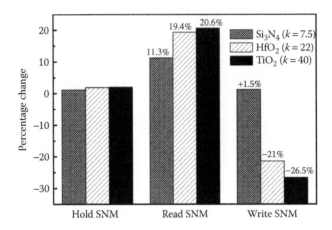

FIGURE 5.7   Percentage change in SNM (hold, read, and write modes of operation) as a function of inner high-k spacer material in the AsymD-kS SRAM cell configuration.

inverters. The access transistor terminal with low-k spacer is connected to the storage node, and the terminal with dual-k is connected to the bit-line.

Compared to the conventional SRAM, this configuration marginally increases the hold SNM by 2%, due to better short-channel characteristics (Bansal et al. 2007). During read mode, the access transistor terminal having dual-k spacer is on high potential, whereas the pull-down transistor terminal with dual-k spacer is at ground. So, the PD_L strength increases more as compared to PA_L; which in turn, increases CR and read stability. The asymmetry between access and pull-down transistors almost doubles the CR that enhances read stability by 19.35%. However, a nearly 20% decrement in write margin is observed due to the increased pull-up transistor strength as compared to the access transistor, which increases PR (even greater than that of the conventional SRAM cell).

Figure 5.7 shows the percentage change in SNM as a function of high-k spacer material. It is observed that with an increase in inner spacer permittivity, both the hold and read SNMs increase; however, the write margin degrades. This reduced write margin can be overcome using an inner spacer with the dielectric value ranging from 7 to 15, but it slightly degrades the

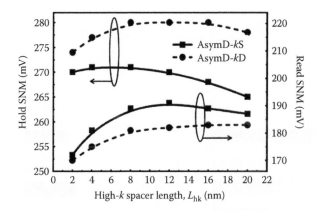

FIGURE 5.8   Effect on hold and read SNMs as a function of L_{hk}.

read SNM as well. Moreover, without affecting the read SNM in an AsymD-*k*S SRAM cell, the write margin can also be improved by using write-assist circuitry. For almost the same read delay, the write delay is substantially reduced by 23.63% due to better drive current of stronger PA_R and PU_R.

Figure 5.8 compares the hold and read SNMs of AsymD-*k* cell configurations. It is observed that both SNMs are at peak for the optimum L_{hk}; and thereafter, they reduce. The asymmetry in terms of I_{ON} between the AsymD-*k*S and AsymD-*k*D structures is maximum at the optimal point ($L_{hk} = 12$ nm) (as observed from Figure 3.20); thus the enhancement in CR and read SNM is maximum at this point.

5.6 EFFECT OF SUPPLY VOLTAGE ON DUAL-*k*–BASED SRAM CELLS

Reducing leakage in SRAM is critical for the overall reduction of the static power consumption in a nanoregime (Giraud et al. 2009). Supply voltage reduction is a technique for lowering leakages, but the reduction in noise immunity limits the supply voltage lowering (Vatajelu and Figueras 2008). Therefore, V_{DD} scalability on SRAM design metrics needs to be explored.

Figure 5.9 presents the effect of lowering the V_{DD} on SNMs (in hold, read, and write modes) and write access time. Compared to

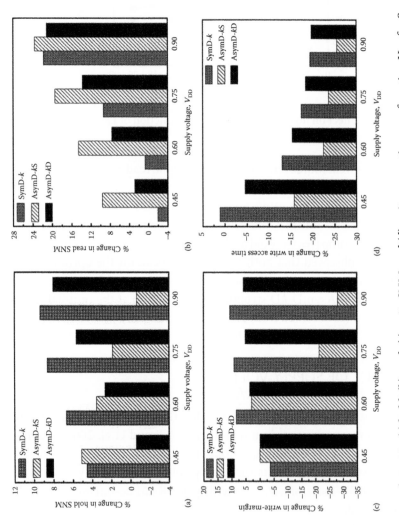

FIGURE 5.9 Percentage change in (a) hold, (b) read, (c) write SNM, and (d) write access time as a function V_{DD} for SymD-k, AsymD-kS, and AsymD-kD SRAM cells with respect to the conventional cell.

the conventional cell, the SymD-k FinFET-based SRAM configuration shows an improvement of 4.5%–9.4% in hold SNM, and up to 21.7% in read stability for supply voltage ranging from 0.45 to 0.9 V. Consequently, the write margin enhances up to 10.6% and write access time reduces by 19.4%. In contrast to this, the AsymD-kD FinFET-based SRAM configuration shows a 2.8%–21.2% improvement in read SNM, 3.6%–5.82% in write margin, a 2.6%–8.0% increase in hold SNM, and a 4.5%–19.7% reduction in write access time. Furthermore, the AsymD-kS–based SRAM demonstrates 5.3%–23.6% higher read SNM and a marginally higher hold SNM at the expense of 15.8%–21.5% write margin. Similar to the inverter VTC (as shown in Figure 4.15), it is observed that the hold SNM in an AsymD-kS cell also reduces with an increase in supply voltage. Conversely, SymD-k and AsymD-kD SRAM cells perform better in the above-threshold region and show significant improvement with higher V_{DD}. The obtained results show that the proposed SymD-k– and AsymD-kD–based SRAM configurations are not a viable candidate in the subthreshold region of operation. The percentage improvement metrics decreases with scaling down of supply voltage. This is due to the reduction in gate fringe coupling through inner high-k spacer that increases series resistance with reduced V_{DD}.

5.7 SYMMETRIC AND ASYMMETRIC DUAL-k SRAM CELL COMPARISON

The robustness of an SRAM cell is commonly evaluated by the SNMs during hold and functional operations of read and write. These three metrics (hold SNM, read SNM, and write SNM) are widely used for design and performance analysis of the SRAM cell.

Figure 5.10 compares the read/write access time and 1-bit SRAM cell leakage power among the conventional, SymD-k–, AsymD-kS–, and AsymD-kD–based SRAM cells. For the SymD-k cell, around 56% and 17% reductions in read and write access times, respectively, are achieved in comparison to the conventional cell, as depicted in Figure 5.10. However, nearly 24% improvement in write access time is observed in AsymD-kS with a marginal loss in read access time by 2%. The write access times in the 6T SRAM cell using the symmetric dual-k architecture are significantly reduced due to their higher I_{ON}/C_{GG}. Primarily, the read access time

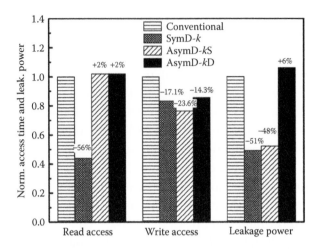

FIGURE 5.10 Comparison of read/write access time and cell leakage power in symmetric and asymmetric dual-k spacer-based SRAM cells normalized with respect to the conventional one.

depends on the read cell current through the access transistors. Therefore, SymD-k outperforms the remaining structures.

Standby leakage power in SRAM is another critical concern that seriously impacts the battery life. The subthreshold current (I_{SUB}) and the gate tunneling (I_G) are considered as the dominant leakage current components to evaluate the total leakage current in an SRAM cell (Bansal et al. 2007). Figure 5.10 shows the estimated leakage power of a 1-bit SRAM cell normalized with respect to the conventional FinFET-based cell. In the OFF-state, I_{SUB} and $I_{EDT-OFF}$ (or, I_{G-OFF}) dominate the total leakage current and both reduce with an increase in L_{un} (Mukhopadhyay et al. 2005). The subthreshold leakage component in SymD-k and AsymD-kS structures is much less than the conventional one because of better electrostatic integrity. Therefore, nearly 50% reduction in cell leakage power is observed for SymD-k– and AsymD-kS–based SRAM cells with respect to the conventional one. Although the AsymD-kD–based SRAM cell mitigates read/write conflict, its total leakage power is almost double that of the AsymD-kS–based SRAM cell. It is due to the higher subthreshold current in the AsymD-kD structure.

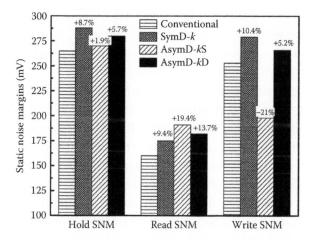

FIGURE 5.11 Comparison of hold SNM, read SNM, and write margin among the conventional, symmetric, and asymmetric dual-*k* spacer-based SRAM cells.

Figure 5.11 compares the SNMs (extracted from the conventional butterfly curves) among the conventional, SymD-*k*-, and AsymD-*k*–based SRAM cells. It is observed that the hold margins in SymD-*k* and AsymD-*k*S SRAMs are increased by 8.6% and 1.2%, respectively, because of the improved drive current. In comparison to the conventional SRAM bit cell, the improvement observed in the read SNM of the AsymD-*k*S–based cell is much higher (19.4%) than the SymD-*k*–based cells (9.4%). This read SNM improvement in the AsymD-*k*S–based SRAM cell is due to its asymmetric nature that helps in adjusting the PR and CR to augment the stability. Contradictorily, the improvement in the SymD-*k* cell is attributed to the enhanced electrostatic integrity that increases SNM without affecting PR and CR. Therefore, it is not justified to directly compare the cell by using only voltage margins, that is, SNMs.

For a suitable comparison among the architectures, it is necessary to explore the architectures on a common metric that clearly distinguishes their superiority in SRAM applications. The N-curve (Wann 2005) metric provides an alternative approach for both current and voltage stability analyses that also satisfies our requirement of comparing SRAM performance on a common platform. As an attractive approach, the N-curve also contains information for both read stability and write stability. The extracted N-curve has three intersection points A, B, and C; wherein, points A and C correspond to stable state points while point B is a meta-stable point (Singh et al. 2013).

The stability metrics derived from the N-curve are based on the combined voltage and current information for an SRAM cell (Singh et al. 2013). The static voltage noise margin (SVNM) is defined as a maximum tolerable DC noise voltage at internal nodes of the cell before its content flips and it is measured as a voltage difference between points B and A. Similarly, static current noise margin (SINM) can be defined as a maximum tolerable DC noise current injected at internal nodes of the cell before its content changes and it is measured as a peak current located between points A and B. These two metrics SVNM and SINM are used to characterize the cell read stability. Similarly, the write stability can be characterized with the help of write-trip voltage (WTV) and write-trip current (WTI). The WTV is the minimum voltage drop needed to change the internal nodes of the cell, which can be measured as a difference between points C and B. The WTI is defined as a minimum amount of the current needed to write the cell, which can be measured as a negative current peak between points C and B. An overlap of points A and B means that the cell is at the edge of stability loss; as a result, destructive read operation can easily occur. Similarly, overlapping of points B and C may lead to failure in the write operation.

It is observed from the N-curves shown in Figure 5.12 that the SymD-k cell shows a better SINM because of high current driving capability and slightly reduced SVNM (due to the same CR) as compared to its asymmetric counterpart AsymD-kS SRAM cell. Contrastingly, the AsymD-kS cell has higher SVNM due to an increase in the CR and lower SINM. Therefore, in the case of these contradictory results, it would be beneficial to derive the power margins (i.e., static power noise margin [SPNM] and write-trip power [WTP]) that include both voltage and current information for read/write stability. The SPNM is defined as the maximum tolerable DC noise power by the internal data storage nodes of a cell before its content changes, and, it is measured as the area under the curve between points A and B. Similarly, the WTP characterizes the write stability of a cell that is measured as the area bounded by the curve and x-axis between points B and C. The inset in Figure 5.12 shows the power margin comparisons of SymD-k– and AsymD-kS–based 6T SRAM cells normalized with respect to the conventional one. For the SymD-k cell, it is observed that the SPNM

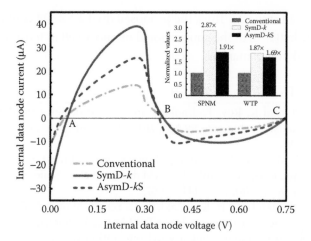

FIGURE 5.12   N-curves and the read/write power margins of conventional, SymD-*k*-, and AsymD-*k*S–based 6T SRAM cells.

and WTP increase by 2.87× and 1.87×, respectively; whereas, the SPNM and WTP increase by 1.91× and 1.69×, respectively, for the AsymD-*k*S–based cell. The percentage change in the SINM with respect to the SVNM is more in the case of SymD-*k* than AsymD-*k*S. Hence, the higher power margins (or area under the curves) are observed in the case of the SymD-*k* cell as compared to the AsymD-*k*S cell. Based on the observations of read/write power margins, the SymD-*k* structure outperforms the AsymD-*k*S for SRAM applications.

5.8 SUMMARY

This chapter investigated the proposed SymD-*k* and AsymD-*k* architectures for high-performance memory application. The proposed dual-*k*–based SRAM cell yields a large reduction in leakage power, improved noise margins, and reduced access time. For the SymD-*k* cell, around 56% and 17% reductions in read and write access times, respectively, are achieved in comparison to the conventional cell. Furthermore, nearly 24% improvement in write access time is observed in AsymD-*k*S with a marginal loss in read access time by 2%. In terms of power margins, the SPNM and WTP in the SymD-*k* cell increase by 2.87× and 1.87×, respectively; whereas, they increase by 1.91× and 1.69×, respectively, in the AsymD-*k*S–based cell. Moreover, there are no cell area penalties associated with the proposed configurations because the device dimensions are the same in all respects. Thus, the proposed dual-k SRAM configurations prove to be better than the conventional memory cell in terms of voltage, current, and power margins.

REFERENCES

Ananthan, H. and K. Roy, Technology and circuit design considerations in quasi-planar double-gate SRAM, *IEEE Trans. Electron Dev.*, 53(2), 242–250, 2006.

Bansal, A., S. Mukhopadhyay, and K. Roy, Device-optimization technique for robust and low-power FinFET SRAM design in NanoScale era, *IEEE Trans. Electron Dev.*, 54(6), 1409–1419, 2007.

Basker, V. S., T. Standaert, H. Kawasaki, C.-C. Yeh, K. Maitra, T. Yamashita, and J. Faltermeier, A 0.063 µm² FinFET SRAM cell demonstration with conventional lithography using a novel integration scheme with aggressively scaled fin and gate pitch, in *2010 Symposium on VLSI Technology*, IEEE, Honolulu, pp. 19–20, 2010.

Bhattacharya, D. and N. K. Jha, FinFETs: From devices to architectures, *Ad. Electron., Hindawi Pub. Corp.*, 2014, 1–21, 2014.

Cakici, R. T. and K. Roy, Analysis of options in double-gate MOS technology: A circuit perspective, *IEEE Trans. Electron Dev.*, 54(12), 3361–3368, 2007.

Chang, C. Y., T.-L. Lee, C. Wann, L.-S. Lai, H.-M. Chen, C.-C. Yeh, and C.-S. Chang, A 25-nm gate-length FinFET transistor module for 32nm node, in *IEDM Technology Digest*, IEEE, Maryland, pp. 293–296, 2009.

Chen, M. C., C.-H. Lin, Y.-F. Hou, Y.-J. Chen, C.-Y. Lin, F.-K. Hsueh, and H.-L. Liu, A 10 nm Si-based bulk FinFETs 6T SRAM with multiple fin heights technology for 25% better static noise margin, in *Proceedings of the IEEE Symposium on VLSI Technology*, IEEE, Kyoto, Japan, pp. T218–T219, 2013.

Cheng, B., S. Roy, and A. Asenov, CMOS 6-T SRAM cell design subject to 'atomistic' fluctuations, *Solid. State. Electron.*, 51(4), 565–571, 2007.

Cheng, H. W. and Y. Li, 16-nm multigate and multifin MOSFET device and SRAM circuits, in *Next-Generation Electronics*, IEEE, Taiwan, pp. 32–35, 2010.

Dobrovolny, P., P. Zuber, M. Miranda, M. G. Bardon, T. Chiarella, P. Buchegger, and K. Mercha, Impact of fin height variations on SRAM yield, in *International Symposium on VLSI Technology, Systems, and Applications*, IEEE, Taiwan, pp. 1–2, 2012.

Ebrahimi, B., R. Asadpour, and A. Kusha, Low-power and robust SRAM cells based on asymmetric FinFET structures, in *4th Asia Symposium on Quality Electronic Design*, IEEE, Penang, Malaysia, pp. 41–45, 2012.

Endo, K., S.-I. O'uchi, Y. Ishikawa, Y. Liu, T. Matsukawa, K. Sakamoto, and M. Masaha, Independent-gate four-terminal FinFET SRAM for drastic leakage current reduction, in *Integrated Circuit Design and Technology and Tutorial*, IEEE, Grenoble, France, pp. 63–66, 2008.

Giraud, B. and A. Amara, Read stability and write ability tradeoff for 6T SRAM cells in double-gate CMOS, in *Proceedings of IEEE International Symposium on Electronic Design, Test and Application*, IEEE, Hong Kong, China, pp. 201–204, 2008.

Giraud, B., O. Thomas, A. Amara, A. Vladimirescu, and M. Belleville, *Planar Double-Gate Transistor: From Technology to Circuit*, chapter 7, Springer, Germany, 2009.

Goel, A., S. K. Gupta, and K. Roy, Asymmetric drain spacer extension (ADSE) FinFETs for low-power and robust SRAMs, *IEEE Trans. Electron Devices*, 58(2), 296–308, 2011.

Guo, Z., S. Balasubramanian, R. Zlatanovici, T.-J. King, and B. Nicolic, FinFET-based SRAM design, in *Proceedings of International Symposium on Low Power Electronics and Design*, IEEE, California, pp. 2–5, 2005.

Gupta, S. K., S. P. Park, and K. Roy, Tri-mode independent-gate FinFETs for dynamic voltage/frequency scalable 6T SRAMs, *IEEE Trans. Electron Devices*, 58(11), 3837–3846, 2011.

Horiguchi, N. et al., High yield sub-0.1m² 6T-SRAM cells, featuring high-k/metal-gate finfet devices, double gate patterning, a novel fin etch strategy, full-field EUV lithography and optimized junction design and layout, in *2010 Symposium on VLSI Technology*, IEEE, Honolulu, pp. 23–24, 2010.

Hu, V. P., M. Fan, P. Su, and C. Chuang, Evaluation of static noise margin and performance of 6T FinFET SRAM Cells with asymmetric gate to source/drain underlap devices, in *IEEE International Silicon on Insulator Conference*, IEEE, San Diego, pp. 1–2, 2010.

International Technology Roadmap for Semiconductors. Available: http://public.itrs.net, 2013.

Joshi, R. V., K. Kim, R. Q. Williams, E. J. Nowak, and C. Te Chuang, A high-performance, low leakage, and stable SRAM row-based back-gate biasing scheme in FinFET technology, in *Proceedings of the IEEE International Conference on VLSI Design*, IEEE, Banglore, India, pp. 665–670, 2007.

Kang, M. S., C. Song, S. H. Woo, H. K. Park, L. Ge, B. M. Han, J. Wang, G. Yeap, and S. O. Jung, FinFET SRAM optimization with fin thickness and surface orientation, *IEEE Trans. Electron Dev.*, 57(11), 2785–2793, 2010.

Kanj, R., R. Joshi, K. Kim, R. Williams, and S. Nassif, Statistical evaluation of split gate opportunities for improved 8T/6T column decoupled SRAM cell yield, in *9th International Symposium on Quality Electronic Design*, IEEE, California, pp. 702–707, 2008.

Kawasaki, H., V. S. Basker, T. Yamashit, C.-H. Lin, Y. Zhu, J. Flatermeier, and S. Schmitz, Challenges and solutions of FinFET integration in an SRAM cell and a logic circuit for 22 nm node and beyond, in *IEDM Technology Digest*, IEEE, Maryland, pp. 1–4, 2009.

Kawasaki, H., M. Khater, M. Guillorn, N. Fuller, J. Chang, S. Kanakasabapathy, and L. Chang, Demonstration of highly scaled FinFET SRAM cells with high-*k* metal gate and investigation of characteristic variability for the 32 nm node and beyond, in *IEDM Technology Digest*, IEEE, California, pp. 1–4, 2008.

Kim, S. H. and J. G. Fossum, Design optimization and performance projections of double-gate FinFETs with gate-source/drain underlap for SRAM application, *IEEE Trans. Electron Dev.*, 54(8), 1934–1942, 2007.

Lee, J. et al., Impact of fin thickness and height on read stability/write ability in tri-gate FinFET based SRAM, in *SOC Design Conference ISOCC*, IEEE, South Korea, pp. 479–482, 2012.

Lekshmanan, D., A. Bansal, and K. Roy, FinFET SRAM: Optimizing silicon fin thickness and fin ratio to improve stability at iso area, in *IEEE Custom Integrated Circuits Conference*, pp. 623–626, 2007.

Liu, Y. et al., Investigation of the TiN gate electrode with tunable work function and its application for FinFET fabrication, *IEEE Trans. Nanotechno.*, 5(6), 723–730, 2006.

Liu, Z., S. A. Tawfik, and V. Kursun, Statistical data stability and leakage evaluation of FinFET SRAM cells with dynamic threshold voltage tuning under process parameter fluctuations, in *Proceedings of the 9th International Symposium on Quality Electronic Design*, IEEE, San Jose, pp. 305–310, 2008.

Masahara, M. et al., Independent double-gate FinFETs with asymmetric gate stacks, *Microelectron. Eng.*, 84(9–10), 2097–2100, 2007.

Matsukawa, T., S. O'uchi, and K. Endo, Comprehensive analysis of variability sources of FinFET characteristics, in *Proceedings of Symposium on VLSI Technology*, Honolulu, HI, IEEE, pp. 118–119, June 2009.

Moradi, F., S. K. Gupta, G. Panagopoulos, D. T. Wisland, H. Mahmoodi, and K. Roy, Asymmetrically doped FinFETs for low-power robust SRAMs, *IEEE Trans. Electron Dev.*, 58(12), 4241–4249, 2011.

Mukhopadhyay, S., K. Kim, C. Chuang, and K. Roy, Modeling and analysis of total leakage currents in nanoscale double gate devices and circuits, in *Proceedings on International Symposium on Low Power Electronics and Design*, IEEE, San Diego, pp. 8–13, August 2005.

Rasouli, S. H., K. Endo, and K. Banerjee, Variability analysis of FinFET-based devices and circuits considering electrical confinement and width quantization, in *International Conference on ComputerAided Design*, IEEE, San Jose, CA, pp. 505–512, 2009.

Rostami, M. and K. Mohanram, Dual-Vth independent-gate FinFETs for low power logic circuits, *IEEE Trans. Comput. Aided Des.*, 30(3), 337–349, March, 2011.

Sachid, A. B. and C. Hu, Denser and more stable SRAM using FinFETs with multiple fin heights, *IEEE Trans. Electron Dev.*, 59(8), 2037–2041, 2012.

Salahuddin, S. M., J. Hailong, and V. Kursun, Characterization of FinFET SRAM cells with asymmetrically gate underlapped bitline access transistors under process parameter fluctuations, in *IEEE International Conference on Electron Devices and Solid-State Circuits*, IEEE, Hong Kong, China, pp. 1–2, 2013a.

Salahuddin, S. M., J. Hailong, and V. Kursun, A novel 6T SRAM cell with asymmetrically gate underlap engineered FinFETs for enhanced read data stability and write ability, in *International Symposium on Quality Electronic Design*, IEEE, California, pp. 353–358, 2013b.

Seevinck, E., F. List, and J. Lohstroh, Static-noise margin analysis of MOS SRAM cells, *J. Solid-State Circuit*, 25(2), 754–784, 1987.

Shimeng, Y., Z. Yuning, D. Gang, and J. Kang, Impact of stochastic mismatch on FinFETs SRAM cell induced by process variation, in *Electron Devices and Solid-State Circuits*, IEEE, Hong Kong, China, pp. 1–4, 2008.

Singh, J., S. P. Mohanty, and D. K. Pradhan, *Robust SRAM Designs and Analysis*, Springer, Germany, 2013.

Sohn, C. W. et al., Effect of fin height of tapered FinFETs on the sub-22-nm System on Chip (SoC) application using TCAD simulation, in *VLSI Technology, Systems, and Applications*, IEEE, Taiwan, pp. 1–2, 2013.

Song, X. et al., Impact of DIBL variability on SRAM static noise margin analyzed by DMA SRAM TEG, in *Technology Digest IEDM*, IEEE, California, pp. 62–65, 2010.

Tawfik, S. A. and V. Kursun, Independent-gate and tied-gate FinFET SRAM Circuits: Design guidelines for reduced area and enhanced stability, in *Proceedings of the International Conference on Microelectronics*, IEEE, Cairo, Egypt, pp. 171–174, 2007.

Tawfik, S. A. and V. Kursun, Portfolio of FinFET memories: Innovative techniques for an emerging technology, in *International SoC Design Conference*, IEEE, South Korea, pp. I-101–I-104, 2008.

Vatajelu, E. I. and J. Figueras, "Supply voltage reduction in SRAMs: Impact on static noise margins," in *Proc. IEEE Int. Conf. Autom. Qual. Testing, Robot*, Romania, vol. 1, pp. 73–78, 2008.

Vega, R. A. and T. J. K. Liu, Comparative study of FinFET versus quasi-planar HTI MOSFET for ultimate scalability, *IEEE Trans. Electron Dev.*, 57(12), 3250–3256, 2010.

Veloso, A. et al., Demonstration of scaled 0.099m² FinFET 6T-SRAM cell using full-field EUV lithography for (Sub-) 22 nm node single patterning technology, in *IEDM Technology Digest*, IEEE, Maryland, pp. 1–4, 2009.

Wann, C., SRAM cell design for stability methodology, in *IEEE International Symposium on VLSI Technology*, IEEE, Hsinchu, China, pp. 21–22, April 2005.

Yang, Y., H. Jeong, F. Yang, J. Wang, G. Yeap, and S. O. Jung, Read-preferred SRAM cell with write-assist circuit using back-gate ETSOI transistors in 22-nm technology, in *IEEE Trans. Electron Dev.*, 59(10), 2575–2581, 2012.

CHAPTER 6

Statistical Variability and Sensitivity Analysis

6.1 INTRODUCTION

Process variability has emerged as one of the major concerns in sub-20 nm gate lengths. Primarily, process variations can be classified as *systematic* and *random*. The systematic variations are predictable in nature and they depend on various deterministic factors such as the layout structure and the surrounding topological environment (Mehrotra et al. 2000; Orshansky et al. 2000). On the other hand, random variations are totally unpredictable and are caused by random uncertainties in the fabrication process such as microscopic fluctuations in the number and location of dopant atoms in the channel region (Tang et al. 1997). Random variations are harder to characterize and can have a negative effect on the yield of critical modules in a design. Random variations can cause a significant mismatch in neighboring devices; and hence, are largely responsible for the poor yield in complex circuits such as SRAMs (Cheng et al. 2004).

As memory will continue to consume a large fraction of the area in future ICs, scaling of memory density must continue to track the scaling trends of logic. Because of the technology scaling, process variations are a critical issue for SRAM stability (Cheng et al. 2004). There are various statistical variability sources for a FinFET-based SRAM that affect the stability, and these issues are required to be addressed. Random discrete dopant fluctuations (RDFs), gate oxide roughness variations (TOX), and metal-grain-dependent work-function variations (WFV) increase the spread in threshold voltage; and thus, the ON- and OFF-currents as the device is scaled down in the nanoscale regime. Increased transistor leakage and parameter variations present the biggest challenges for the scaling of six-transistor (6T) SRAM memory arrays. Therefore, for an optimized FinFET SRAM design, solutions are needed to address these issues. Moreover, the control of key structural dimensions such as L_G, T_{fin}, and so on continues to be very difficult, due to technological (lithographic and etching) restrictions.

The significance of random variability and sensitivity of key structural parameters in nanoscaled tri-gate transistors increases sharply with the downscaling of device dimensions below 20 nm gate lengths that can cause failure in any design. Therefore, it is necessary to investigate the proposed dual-k devices and their circuit performance under process variations. This chapter briefly presents a comparison among the conventional, SymD-k, and AsymD-kS tri-gate architectures based on device/SRAM parameters such as V_{th}, I_{ON}, I_{OFF}, and SNMs under random (specifically RDF, TOX, WFV) and parametric (T_{fin}, L_{hk}, dopant segregation length [DSL]) variations.

6.2 IMPEDANCE FIELD METHOD (IFM) AND TECHNOLOGY COMPUTER-AIDED DESIGN (TCAD) SIMULATION SETUP

As transistor scaling continues, self-averaging of device properties for individual devices becomes less effective; and therefore, the statistical variability of device properties becomes more prominent. For the investigation of variability in single transistors, so-called atomistic approaches have been proposed to investigate the variability for MOSFETs (Cheng et al. 2007). Such methods rely on 3D TCAD simulations of a large number of independent randomized 3D realizations of the device structure. The computational resources needed for the "atomistic" approach are directly proportional to the number of randomized device structures in the statistical sample. Such an approach is therefore naturally limited to smaller devices with simplified geometries (Cheng et al. 2007).

The application of atomistic methods to 3D 6T SRAM cells with realistic geometries seems to be extremely difficult. For the exploration of random variability in SRAM cells, quite a large number of theoretical- and simulation-based approaches were reported (Bhavnagarwala et al. 2001; Hiramoto et al. 2011; Hu et al. 2011). In Bhavnagarwala et al. (2001), analytic approximations are used. In Hiramoto et al. (2011), the SRAM cell is modeled using the Simulation Program with Integrated Circuit

Emphasis (SPICE) tool, wherein transistors are represented as compact models. In Hu et al. (2011), 3D TCAD simulations for an entire SRAM cell are presented, where, the cell geometric and doping profiles are defined analytically.

Recently, the statistical IFM has been reported as a viable alternative to atomistic and SPICE-based approaches. The IFM in TCAD provides a fast, convenient, and accurate alternative for statistical variability analysis (Sayed et al. 2012b). The basic concept behind the IFM is to treat randomness as a perturbation of a reference device. Rather than solving the full nonlinear Poisson and drift-diffusion equations for a large number of random device realizations, the 3D TCAD solution is obtained only once for the reference device. Then, the current fluctuations at the device terminals caused by these random perturbations are computed. These computations are based on linear response theory using Green's function technique (Bonani et al. 1998; Wettstein et al. 2003). The IFM can be applied to different kinds of perturbations, including geometric fluctuations and work-function fluctuations. The most prominent advantage of the IFM method is that, it is applicable to large device structures and can also readily handle realistic geometries. Statistical IFM can also be used for more complex applications, such as the investigation of the static noise margin variability of SRAM cells (Sayed et al. 2012a). The computational resource requirements depend weakly on the number of randomized devices included in the statistical sample.

The random device-circuit variability of the proposed SymD-k and AsymD-kS architectures is investigated in terms of random discrete dopant (RDF)-induced variations, gate oxide roughness variations (TOX), and metal grain-dependent WFV. The doping profile is randomized according to Sano's method (Sano et al. 2002). To obtain statistical samples, the doping is spatially uncorrelated and that the number of dopants in a given volume follows a Poisson distribution, with an average number of dopants in the volume.

To study the WFV, gate metals TiN and MoN are used for n-type and p-type, respectively (Rasouli et al. 2014). TiN metal exhibits random positional dependence of two possible grain orientations (<200> and <111>) with work-function values of 4.6 and 4.4 eV, and the probabilities of occurrences of 60% and 40%, respectively. Similarly, MoN exhibits 5.0 and 4.4 eV for grain orientations of <110> and <112>, with 60% and 40% probabilities of occurrences, respectively. The average grain sizes for TiN and MoN metals are 22 nm and 17 nm, respectively. For gate oxide roughness variations (TOX), a procedure similar to the WFV is used. Here, in each "grain" along the surface, the oxide thickness is modified. In this chapter, all the random variability simulations using IFM are carried out at the supply voltage of 1 V.

6.3 RESULTS AND DISCUSSIONS

This section briefly presents the comparison among conventional, SymD-k, and AsymD-kS FinFET architectures based on the performance parameters such as I_{ON}, I_{OFF}, V_{th}, inverter delay, and SRAM static noise margin (SNM) under random statistical and parametric variations.

6.3.1 Statistical Variability of SymD-k and AsymD-kS Structures

The random device-circuit variability of both the SymD-k device and the conventional device is investigated in terms of RDF, TOX, and metal grain-dependent WFV. The numerical comparison of the standard deviations and the relative variation for V_{th}, I_{ON}, and I_{OFF} is summarized in Table 6.1 and their total impact is shown in Figure 6.1.

It is observed that both SymD-k and AsymD-kS structures are more immune to random variations as compared to their conventional counterpart. As the channel and underlap region is lightly doped in all considered structures, a reduced threshold (V_{th}) fluctuation, that is, ~1.6%–1.9%, is observed under RDF-induced variations. The σV_{th} values due to oxide thickness variations in SymD-k and AsymD-kS are observed as 1.84 and 4.11 mV, respectively, in comparison to 6.39 mV of the conventional device. However, the effect of work function on σV_{th} is significantly higher due to the enhanced grain-oriented work-function difference. Overall, the SymD-k architecture demonstrates reduced V_{th} variations to their mean values due to its superior electrostatic and enhanced gate controllability over the channel in comparison to the AsymD-kS and conventional devices.

On the other hand, it is observed that the AsymD-kS device outperforms the other architectures in terms of I_{ON} variation. In the AsymD-kS device, an ON-current deviation with RDF is relatively lower than the other architectures. It is because of the highly concentrated carriers near the G/S surface and enhanced fringing field; that in turn, establishes immunity to I_{ON}. It is observed that the current distribution (σI_{ON}) and percentage variation ($\sigma I_{ON}/<I_{ON}>$) in the ON-state are less affected due to TOX and WFV as compared to RDF variations. Contrastingly, σI_{OFF} and ($\sigma I_{OFF}/<I_{OFF}>$) are highly affected by TOX and WFV as compared to RDF variations.

TABLE 6.1 Standard Deviation and Its Relative Variation (in %) of Device Parameters in Conventional, SymD-k, and AsymD-kS Tri-gate FinFET Architectures

		V_{th} (mV)			I_{ON} (µA)			I_{OFF} (pA)		
		Conv.	SymD-k	AsymD-kS	Conv.	SymD-k	AsymD-kS	Conv.	SymD-k	AsymD-kS
RDF	σ	4.61	4.38	4.11	1.60	2.50	1.56	8.93	6.48	4.62
	%	1.91	1.60	1.65	8.58	5.73	4.23	11.01	12.21	11.66
TOX	σ	6.39	1.84	4.11	0.82	1.45	0.79	34.29	7.75	15.44
	%	2.65	0.67	1.65	4.40	3.33	2.14	42.39	20.52	29.49
WFV	σ	17.95	16.10	17.70	0.38	0.73	0.38	37.68	16.98	24.20
	%	7.43	5.87	7.12	2.06	1.68	1.04	46.36	44.86	46.22
Total	σ	19.58	16.79	18.67	2.14	2.99	1.79	45.87	18.12	26.96
	%	8.11	6.12	7.51	9.67	6.85	4.86	56.69	47.89	51.61

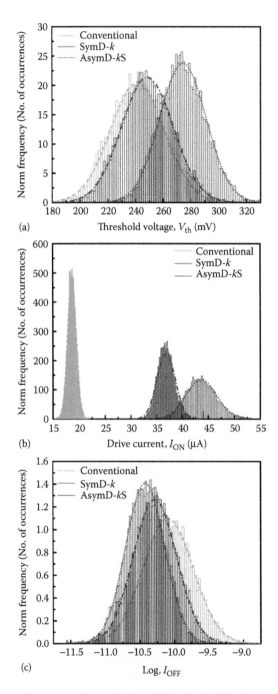

FIGURE 6.1 Total statistical variability of (a) threshold voltage, (b) drive current, and (c) Log I_{OFF} of conventional, SymD-k, and AsymD-kS considering RDF, TOX, and WFV.

The percentage variations due to TOX and WFV obtained in I_{OFF} for the SymD-k structure are much better than the AsymD-kS, and its mean value ($<\sigma I_{OFF}>$) is more than two times lower than the conventional one. Percentage variations of all parameters are found to be much lower (except for a marginally higher I_{ON}) in a SymD-k–based device compared to the AsymD-kS and conventional devices. The WFV-induced variations in I_{ON} are comparable for both the devices, given that the mean value ($<I_{ON}>$) for SymD-k is 2.4x higher than the conventional one.

6.3.2 Statistical Variability of SymD-k and AsymD-kS–Based SRAM Cells

There is an emerging need of robust and high-performance SRAMs, since, memories occupy most of the die area in processors (ITRS 2013). Semiconductor memories have a large number of transistors; wherein, a small variation in device parameter leads to failure of read/write operations. Variations in device dimensions can severely affect the balance of transistor ratio that often degrades the read/write stability. This subsection describes the design metric outcome and statistical variations of SymD-k and AsymD-kS FinFET-based SRAM cells in comparison to the conventional one.

From SRAM perspectives, both the SymD-k– and AsymD-kS–based 6T cells yield improved noise margins and reduced access times, without affecting the cell area. Compared to the conventional SRAM cell, the SymD-k cell improves read SNM by 5%, 9.4%, and 10.6% for spacer material of Si_3N_4, HfO_2, and TiO_2, respectively. Moreover, the write-access time substantially reduces by 4.4%, 17.1%, and 20.7% for spacer material of Si_3N_4, HfO_2, and TiO_2, respectively. It is due to the improved current/capacitive behavior at an optimum L_{hk} of 12 nm. As the read-access time primarily depends on the current driving capability of an access transistor, a large improvement of ~24%–64% is observed for spacer permittivity ranging from 7.5 to 40.

The improvement observed in SymD-k and AsymD-kS device parameters; and their relative variations, in turn, improve SRAM variability also. The random variability effects on hold and read SNMs are shown in Figures 6.2 and 6.3, respectively, using conventional, SymD-k, and AsymD-kS structures. Moreover, the relative variations (in %) and standard deviations for hold and read SNMs are shown in Table 6.2. The relative variations in SRAM performances are found to be much lower using a SymD-k device in comparison to the conventional FinFET.

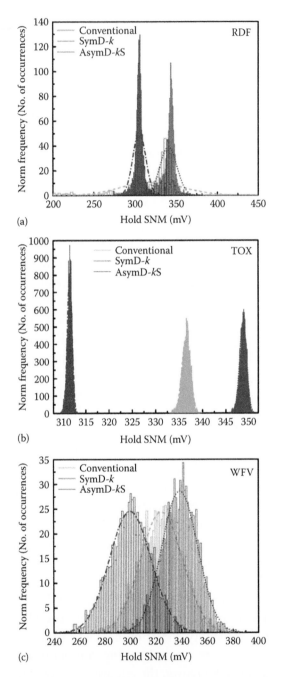

FIGURE 6.2 Statistical variability of the hold SNM of conventional, SymD-k–, and AsymD-kS–based 6T SRAM cells as a function of (a) RDF, (b) TOX, and (c) WFV.

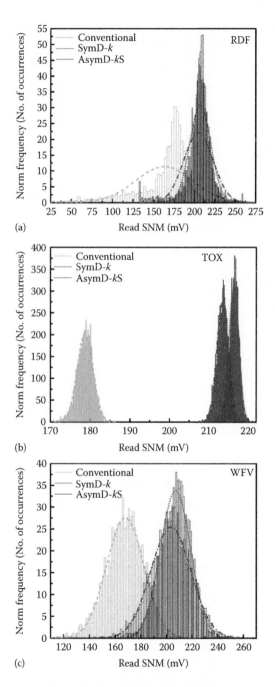

FIGURE 6.3 Statistical variability of the read SNM of conventional, SymD-*k*–, and AsymD-*k*S–based 6T SRAM cells as a function of (a) RDF, (b) TOX, and (c) WFV.

TABLE 6.2   Standard Deviation and Its Relative Variation of Hold/Read SNM in Conventional, SymD-k–, and AsymD-kS–based SRAM Cells

		Hold SNM			Read SNM		
		Conv.	SymD-k	AsymD-kS	Conv.	SymD-k	AsymD-kS
RDF	Std. dev (mV)	37.07	10.16	7.73	35.16	12.94	17.97
	% variations	11.58	2.98	2.53	21.55	6.25	8.77
TOX	Std. dev (mV)	0.90	0.75	0.50	1.97	1.27	1.45
	% variations	0.27	0.22	0.16	1.11	0.59	0.68
WFV	Std. dev (mV)	16.39	13.93	16.28	14.50	11.82	15.67
	% variations	5.06	4.12	5.43	8.65	5.71	7.71

6.4 SENSITIVITY ANALYSIS

The sensitivity analysis provides a relative significance of each device parameter on the performance metrics. The sensitivity metric on a parameter (p) is defined as

$$S(M) = 100 \times \left[\frac{(\delta M/M)}{(\delta p/p)} \right]$$

where, M is the performance parameter that depends on three device parameters (p), such as silicon film thickness (T_{fin}), inner high-k spacer length (L_{hk}), and DSL.

Table 6.3 compares the performance of SymD-k and AsymD-kS devices with the conventional one by considering ±20% variation in key structural parameters. It is observed that the inverter delay in the SymD-k (AsymD-kS) structure is nearly 2.1× (2.5×) and 1.34× (1.63×) lesser sensitive to the T_{fin}

TABLE 6.3   Sensitivity (in Percentage) of Conventional, SymD-k, and AsymD-kS Structure Considering ±20% Structural variations

	Conv.		SymD-k			AsymD-kS		
Sensitivity (%)	T_{fin}	DSL	T_{fin}	DSL	L_{hk}	T_{fin}	DSL	L_{hk}
V_{th}	28	5	14	2	2	20.9	4	1.9
I_{ON}	44	67	23	64	44	37.5	56.2	26.9
I_{OFF}	560	76	368	58	32	448.3	72.8	25.5
C_{GG}	36	14	17	23	64	18	27	57.5
SS	20	5	14	4	1	15.2	4.98	0.7
Inverter delay	19	31	9	23	11	7.6	19	13.3

and DSL variation, respectively, in comparison to its conventional counterpart. Both the SymD-k and AsymD-kS inherently introduce an additional source of structural sensitivity, that is, inner high-k spacer length (L_{hk}), that actually favors the charge dynamics in the channel. Due to an optimized L_{hk}, the fringe capacitance component is much higher in the case of dual-k; consequently, the overall gate capacitance is more responsive to L_{hk}. In both the proposed dual-k architectures, the optimization of L_{hk} depends on DSL; therefore, the inverter delay is more sensitive to L_{hk} and DSL as compared to T_{fin}. The silicon film thickness marginally affects the dynamic performance of the circuit while it highly amends the short-channel characteristics such as I_{OFF} and subthreshold slope.

6.5 SUMMARY

This chapter analyzed the random variability and structural sensitivity of the proposed SymD-k and AsymD-kS architectures. In comparison to the conventional one, the SymD-k (AsymD-kS) architecture improves the random variations of V_{th}, I_{ON}, and I_{OFF} by 24.5% (7.39%), 29.16% (49.74%), and 15.5% (8.96%), respectively. Furthermore, the SymD-k–based SRAM cell also exhibits more immunity to hold and read SNMs. It is also observed that the inverter delay in the SymD-k structure shows lesser sensitivity, nearly 2.1× (2.5×) and 1.34× (1.63×) to the T_{fin} and DSL, respectively. Overall, both dual-k architectures exhibit the least sensitivity to random variations in comparison to their conventional counterpart, which proves them to be the suitable candidates for high-performance device/circuit applications in sub-20 nm nodes.

REFERENCES

Bhavnagarwala, A. J., X. Tang, and J. D. Meindl, The impact of intrinsic device fluctuations on CMOS SRAM cell stability, *IEEE J. Solid State Circuits*, 36(4), 658–665, 2001.

Bonani, F., G. Ghione, M. R. Pinto, and R. K. Smith, An efficient approach to noise analysis through multidimensional physics-based models, *IEEE Trans. Electron Dev.*, 45(1), 261–269, 1998.

Cheng, B., S. Roy, and A. Asenov, The impact of random doping effects on CMOS SRAM cell, in *Proceedings of European Solid-State Circuits Conference*, IEEE, Belgium, Germany, pp. 219–222, 2004.

Cheng, B., S. Roy, and A. Asenov, CMOS 6-T SRAM cell design subject to 'atomistic' fluctuations, *Solid. State. Electron.*, 51(4), 565–571, 2007.

Hiramoto, T., M. Suzuki, X. Song, K. Shimizu, T. Saraya, A. Nishida, and T. Tsunomura, Direct measurement of correlation between SRAM noise margin and individual cell transistor variability by using device matrix array, *IEEE Trans. Electron Dev.*, 58(8), 2249–2256, 2011.

Hu, V. P., M. Fan, C. Hsieh, P. Su, and C. Chuang, FinFET SRAM cell optimization considering temporal variability due to NBTI/PBTI, surface orientation and various gate dielectrics, *IEEE Trans. Electron Dev.*, 58(3), 805–811, 2011.

International Technology Roadmap for Semiconductors. Available: http://public. itrs.net, 2013.

Mehrotra, V., S. L. Sam, D. Boning, A. Chandrakasan, R. Vallishayee, and S. Nassif, A methodology for modeling the effects of systematic within-die interconnect and device variation on circuit performance, in *Proceedings of Design Automation Conference*, IEEE, California, pp. 172–175, 2000.

Orshansky, M., L. Milor, P. Chen, K. Keutzer, and C. Hu, Impact of systematic spatial intra-chip gate length variability on performance of high-speed digital circuits, in *Proceedings of International Conference on Computer-Aided Design*, IEEE, pp. 62–67, 2000.

Rasouli, S. H., K. Endo, J. F. Chen, N. Singh, and K. Banerjee, Grain-oriented induced quantum confinement variation in FinFETs and multi-gate ultra-thin body CMOS devices and implications for digital design, *IEEE Trans. Electron Dev.*, 58(8), 2282–2291, 2014.

Sano, N., K. Matsuzawa, M. Mukai, and N. Nakayama, On discrete random dopant modeling in drift-diffusion simulations: Physical meaning of atomistic dopants, *Microelectron. Rel.*, 42(2), 189–199, 2002.

Sayed, K. E., E. Lyumkis, and A. Wettstein, Modeling statistical variability with the impedance field method, in *International Conference on Simulation of Semiconductor Processes and Devices*, IEEE, Denver, CO, pp. 205–208, September 2012a.

Sayed, K. E., A. Wettstein, S. D. Simeonov, E. Lyumkis, and B. Polsky, Investigation of the statistical variability of static noise margins of SRAM cells using the statistical impedance field method, *IEEE Trans. Electron Dev.*, 59(6), 1738–1744, 2012b.

Tang, X., V. De, and J. Meindl, Intrinsic MOSFET parameter placement due to random placement, *IEEE Trans. VLSI Syst.*, 5(4), 369–376, 1997.

Wettstein, A. et al., Random dopant fluctuation modelling with the impedance field method, in *International Conference on Simulation of Semiconductor Processes and Devices*, IEEE, San Jose, CA, pp. 91–94, September 2003.

Index

Printed and bound by CPI Group (UK) Ltd, Croydon, CR0 4YY

24/10/2024

01778301-0020